Schriftenreihe der Institute für Systemdynamik (ISD) und optische Systeme (IOS)

Chefredakteure

Jürgen Freudenberger, Institut für Systemdynamik, Hochschule Konstanz (HTWG), Konstanz, Baden-Württemberg, Deutschland

Johannes Reuter, Institut für Systemdynamik, Hochschule Konstanz (HTWG), Konstanz, Baden-Württemberg, Deutschland

Matthias Franz, Institut für Optische Systeme, Hochschule Konstanz (HTWG), Konstanz, Baden-Württemberg, Deutschland

Georg Umlauf, Institut für Optische Systeme, Hochschule Konstanz (HTWG), Konstanz, Baden-Württemberg, Deutschland

Die „Schriftenreihe der Institute für Systemdynamik (ISD) und optische Systeme (IOS)“ deckt ein breites Themenspektrum ab: von angewandter Informatik bis zu Ingenieurswissenschaften. Die Institute für Systemdynamik und optische Systeme bilden gemeinsam einen Forschungsschwerpunkt der Hochschule Konstanz. Die Forschungsprogramme der beiden Institute umfassen informations- und regelungstechnische Fragestellungen sowie kognitive und bildgebende Systeme. Das Bindeglied ist dabei der Systemgedanke mit systemtechnischer Herangehensweise und damit verbunden die Suche nach Methoden zur Lösung interdisziplinärer, komplexer Probleme. In der Schriftenreihe werden Forschungsergebnisse in Form von Dissertationen veröffentlicht.

The "Series of the institutes of System Dynamics (ISD) and Optical Systems (IOS)" covers a broad range of topics: from applied computer science to engineering. The institutes of System Dynamics and Optical Systems form a research focus of the HTWG Konstanz. The research programs of both institutes cover problems in information technology and control engineering as well as cognitive and imaging systems. The connective link is the system concept and the systems engineering approach, i. e. the search for methods and solutions of interdisciplinary, complex problems. The series publishes research results in the form of dissertations.

Weitere Bände in der Reihe https://link.springer.com/bookseries/16265

Daniel Benjamin Rohweder

Signal Constellations with Algebraic Properties and their Application in Spatial Modulation Transmission Schemes

Springer Vieweg

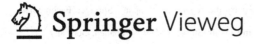

Daniel Benjamin Rohweder
Uhldingen-Mühlhofen, Germany

Dissertation, Ulm University, 2021
Funded by the Deutsche Forschungsgemeinschaft (DFG, German Research Foundation) – FR 2673/6–1
In reference to IEEE copyrighted material which is used with permission in this thesis, the IEEE does not endorse any of the university of Ulm's or Springer Vieweg's products or services.

ISSN 2661-8087 ISSN 2661-8095 (electronic)
Schriftenreihe der Institute für Systemdynamik (ISD) und optische Systeme (IOS)
ISBN 978-3-658-37113-5 ISBN 978-3-658-37114-2 (eBook)
https://doi.org/10.1007/978-3-658-37114-2

Responsible Editor: Stefanie Eggert
This Springer Vieweg imprint is published by the registered company Springer Fachmedien Wiesbaden GmbH part of Springer Nature.
The registered company address is: Abraham-Lincoln-Str. 46, 65189 Wiesbaden, Germany

Acknowledgments

First, I would like to thank my doctoral supervisor Prof. Dr.-Ing. Jürgen Freudenberger for giving me the opportunity to join his research group at the HTWG Konstanz. I am very greatful for the numerous discussions we have had, his support and everything I have learned from him during the last years. It was a pleasure to work with him.

Moreover, I want to thank the German Research Foundation (Deutsche Forschungsgemeinschaft DFG) for funding my work and position at the HTWG Konstanz. I thank my coauthors Dr.-Ing. Sebastian Stern and Prof. Dr.-Ing. Robert Fischer from Ulm in Germany and Prof. Dr. Sergo Shavgulidze from Tbilisi in Georgia. The received feedback was greatly appreciated and always improved the quality of our publications.

Finally, i am deeply greatful to my family for their encouragement. Most of all, I am greatful to my beautiful fiancée Gladys for her love, patience, and support.

Abstract

Nowadays, most digital modulation schemes are based on conventional signal constellations that have no algebraic group, ring, or field properties, e.g. square quadrature-amplitude modulation constellations. Signal constellations with algebraic structure can enhance the system performance. For instance, multidimensional signal constellations based on dense lattices can achieve performance gains due to the dense packing. The algebraic structure enables low-complexity decoding and detection schemes. In this work, signal constellations with algebraic properties and their application in spatial modulation transmission schemes are investigated.

Spatial modulation is a wireless multiple-input multiple-output transmission technique, where not all transmit antennas are active simultaneously. Information is transmitted via two data streams, one defines the subset of active transmit antennas and the other is mapped to a symbol of a signal constellation. The detection complexity at the receiver increases exponentially with the number of active antennas when maximum-likelihood (ML) detection is used. In order to reduce the computational complexity, suboptimal detection methods are necessary.

We investigate two- and four-dimensional signal constellations based on Gaussian, Eisenstein, and Hurwitz integers for the application in spatial modulation transmission. The Eisenstein and Hurwitz constellations are derived from the densest two- and four-dimensional lattices. We propose detection algorithms with reduced complexity. The proposed Eisenstein and Hurwitz constellations combined with the proposed suboptimal detection can outperform conventional two-dimensional constellations with ML detection.

Contents

Notation

Abbreviations

AWGN	additive white Gaussian noise
CSI	channel state information
GAM	Gaussian approximation method
GMSM	generalized multistream spatial modulation
GSM	generalized spatial modulation
LDPC	low-density parity-check
MAC	multiply-accumulate
MGAM	modified Gaussian approximation method
MIMO	multiple-input/multiple-output
ML	maximum likelihood
MMSE	minimum mean square error
MRC	maximum-ratio combiner
MSM	multistream spatial modulation
OMEC	one-Mannheim error-correcting
PDF	probability density function
PSK	phase-shift keying
QAM	quadrature-amplitude modulation
SD	suboptimal detection
SER	symbol error ratio
SISO	single-input/single-output
SM	spatial modulation
SNR	signal to noise ratio
SPM	spatial permutation modulation

STB	space time block
SVD	signal-vector based detection
V-BLAST	vertical-Bell laboratories layered space-time

System Parameters

$\boldsymbol{H}$	$N_{rx} \times N_{tx}$ MIMO channel matrix
$\boldsymbol{H}_i$	$N_{rx} \times N_a$ submatrix of $\boldsymbol{H}$
$\boldsymbol{y}$	receive vector
$\boldsymbol{x}$	signal vector
$\boldsymbol{n}$	noise vector
$\boldsymbol{s}$	symbol vector
$\boldsymbol{e}$	error vector
$\boldsymbol{Y}$	receive matrix (SPM)
$\boldsymbol{X}$	symbol matrix (SPM)
$\boldsymbol{N}$	noise matrix (SPM)
N_{tx}	number of transmit antennas
N_{rx}	number of receive antennas
N_a	number of active antennas
N_p	number of antenna patterns
ρ	signal to noise ratio (all SM transmission scenarious)
E_s	average energy per transmitted symbol
E_b	average energy per information bit
N_c	code length
$\boldsymbol{P}$	parity-check matrix
R	code rate

Number Sets and Symbols

$\mathbb{C}$	the complex numbers
$\mathbb{F}_p$	finite field of order$p$
$\mathbb{H}$	the quaternions
$\mathbb{R}$	the real numbers
$\mathbb{R}^D$	D-dimensional Euclidean space
$\mathbb{Z}$	the integers
$\mathcal{C}$	set of codewords
$\mathscr{E}$	the Eisenstein integers

$\mathcal{E}$	finite set of Eisenstein integers
$\mathscr{G}$	the Gaussian integers
$\mathcal{G}$	finite set of Gaussian integers
$\mathscr{H}$	the Hurwitz integers
$\mathcal{H}$	finite set of Hurwitz integers
$\mathscr{L}$	the Lipschitz integers
$\mathcal{L}$	finite set of Lipschitz integers
$\mathcal{V}$	the set of half integers
$\mathcal{T}$	two-dimensional signal constellation
$\mathcal{F}$	four-dimensional signal constellation
$\mathcal{X}$	multidimensional signal constellation
$\mathcal{P}$	set of SPM transmission pattern
C	channel capacity
μ	expected value
ϵ	channel error probability
i, j, k	imaginary units
M_2	cardinality of a two-dimensional signal constellation
M_4	cardinality of a four-dimensional signal constellation
M_{D}	cardinality of a D-dimensional signal constellation
bpcu	bits per channel use
bpsq	bits per sequence (SPM)
bits_2	bits per two dimensions, e.g. $M_4 = 256$ equals 4 bits_2
ω	complex root of unity $\frac{1}{2}\left(-1+\sqrt{3}\mathrm{i}\right)$
$\mathrm{dist}_{\mathrm{E}}$	Euclidean distance
$\mathrm{dist}_{\mathrm{H}}$	Hamming distance
$\mathrm{dist}_{\mathrm{M}}$	Mannheim distance
δ_2	minimum squared Euclidean distance

Math Operators and Operations

$\mathrm{E}\{\cdot\}$	expected value operator
$\mathrm{CFM}(\cdot)$	constellation figure of merit
$\mathrm{P}(\cdot)$	probability for the occurence of given event
$H(\cdot)$	entropy function
$N(\cdot)$	norm function
$\exp(\cdot)$	natural exponential function
$\mathcal{O}(\cdot)$	big O notation

$\mathrm{wt}(\cdot)$	weight function
$\mathrm{wt}_{\mathrm{M}}(\cdot)$	Mannheim weight
$\lvert\cdot\rvert$	absolute value
$\lfloor\cdot\rfloor$	floor function (rounds to nearest integer less than or equal to it)
$\lceil\cdot\rceil$	ceiling function (rounds to nearest integer greater than or equal to it)
$\lfloor\cdot\rceil$	rounding to closest Gaussian integer
$[\cdot]$	rounding to closest Eisenstein integer
$\boldsymbol{a}$	column vector
$\boldsymbol{A}$	matrix
$\boldsymbol{A}^{\mathsf{T}}$	transpose of matrix $\boldsymbol{A}$
$\boldsymbol{A}^{\mathsf{H}}$	Hermitian transpose of matrix $\boldsymbol{A}$
$\boldsymbol{a}^*$	complex conjugate of $\boldsymbol{a}$
$\boldsymbol{A}^{-1}$	inverse of matrix $\boldsymbol{A}$
$\boldsymbol{I}_m$	$m \times m$ identity matrix
$\mathrm{argmin}\{\cdot\}$	arguments of the minima
$\mathrm{argmax}\{\cdot\}$	arguments of the maxima
$\mathrm{argsort}\{\cdot\}$	arguments in sorted order

Introduction 1

During recent decades, mobile communication systems have had to handle an ever-increasing demand on data throughput, e.g. for subscription-based streaming services, social media platforms, and cross-device file synchronisation via cloud storage. Hence, spectral efficiency and energy efficiency hold strong importance for the design of future mobile communication systems.

For a given wireless communication system, the system performance strongly depends on the signal constellation applied. Nowadays, most digital modulation schemes are based on conventional two-dimensional signal constellations, which are the equivalent complex-baseband representation of radio-frequency signals, e.g. quadrature-amplitude modulation (QAM) or phase-shift keying (PSK).

Usually, conventional signal constellations have no algebraic group, ring, or field properties. Signal constellations that have algebraic group properties enable constructing error-correcting codes over complex-valued alphabets and the algebraic structure can be used for low-complexity detection/decoding schemes. Two-dimensional signal constellations with algebraic group, ring, or field properties were investigated in [Loe91] for PSK, in [Hub94b, FGS13] for Gaussian integers, and in [Hub94a, FBW14, THBN15, FS17] for Eisenstein integers, respectively. Gaussian integers are complex numbers whose real and imaginary parts are both integers. Eisenstein integers are the isomorphic complex-valued representation of the hexagonal lattice, whereas Gaussian integers are isomorphic to the two-dimensional integer lattice. Due to their algebraic properties, many well-known coding techniques for linear codes, e.g. the Plotkin construction or construction of product codes, can be applied to codes over such signal constellations [PIF+01, MBG07, KMI+10, FGS13, SYHS13, FBW14, RFS18].

D. B. Rohweder, *Signal Constellations with Algebraic Properties and their Application in Spatial Modulation Transmission Schemes*, Schriftenreihe der Institute für Systemdynamik (ISD) und optische Systeme (IOS), https://doi.org/10.1007/978-3-658-37114-2_1

With wireless single-input/single-output (SISO) communication systems, i.e. systems with one transmit and one receive antenna, the transmission is typically based on the transmission of signals from a two-dimensional or complex-valued signal constellation. Multidimensional signal constellations are suitable for wireless communication systems, where—in addition to the in-phase (I)- and quadrature (Q)-components—the horizontal and vertical polarization is exploited, as well as for systems with multiple active transmit antennas. Signal constellations of higher dimension can either be created with the Cartesian product of conventional signal constellations or based on a finite set of points, taken from a multidimensional lattice.

It was demonstrated in [FW89, PDL+90, PC93] that multidimensional signal constellations have advantages for multilevel coding compared with two-dimensional signal constellations. Similarly, multidimensional signal constellations that have group properties can improve the performance over the AWGN channel compared with two-dimensional constellations, as demonstrated in [FS15b] for Lipschitz integers. Lipschitz integers form a four-dimensional integer lattice and can be considered as the extension of Gaussian integers to four dimensions. Hurwitz integers are a superset of the Lipschitz integers, where the set additionaly includes the set of half-integers, i.e. Hurwitz integers are the union of the Lipschitz integers and the same set with an 0.5 offset per dimension. The Hurwitz integers are the isomorphic quaternion-valued representation of the D_4 lattice, which is known to be the densest packing in four dimensions [CS99]. Four-dimensional signal sets are important for optical communications, where the transmission is based on the I- and Q-components of both polarizations [KA10]. Moreover, such constellations are useful for broadcast channels [NHV15, Hua17] and multiple-input/multiple-output (MIMO) transmission [KRAK12, FRS18, SF18, USS19, USS20].

In this work, we focus on MIMO transmission based on spatial modulation (SM). SM is a wireless MIMO transmission technique that uses only a single active transmit antenna in each time slot [MHS+08, JGS08, YDX+15]. With generalized multistream spatial modulation (GMSM), a subset of the transmit antennas is used for each transmitted symbol [YSMH10, NDPNV13]. In order to improve the spectral efficiency of SM, the selection of the subset is used to transmit information, i.e. information is transmitted by choosing symbols from a signal constellation and additionally by choosing a subset of the transmit antennas. With GMSM, at least two antennas are active simultaneously. They transmit independent data symbols over each active transmit antenna. At the receiver, maximum-likelihood (ML) detection may be used to retrieve the transmitted symbol vector. For high spectral efficiency or low-latency applications, ML detection is not feasable and suboptimal detection techniques are necessary.

In this thesis, we consider and analyze signal constellations that have algebraic properties. We apply multidimensional signal constellations to SM transmission schemes and exploit the algebraic structure for low-complexity decoding and detection methods. For instance, Gaussian- and Eisenstein-integer constellations enable constructing short and yet powerful codes that can be used as a multidimensional signal constellation. Their algebraic properties enable a low-complexity list decoding algorithm that achieves near-ML performance. Like the two-dimensional case, multidimensional constellations can be constructed based on lattices. These lattices can obtain dense sphere packings and have an algebraic structure.

The remainder of this thesis is organized as follows:

Chapter 2 provides an introduction to the channel models considered and SM transmission scenarios that are used througout this work. It includes the additive white Gaussian noise (AWGN) channel, as well as the Rayleigh fading channel. The SM-MIMO transmission schemes, SM [MHAY06], generalized spatial modulation (GSM) [YSMH10], GMSM [WJS12a], and spatial permutation modulation (SPM) [LSL+19] are briefly reviewed.

Chapter 3 focuses on two- and four-dimensional signal sets that have algebraic properties. Codes over Gaussian- and Eisenstein-integer fields, so-called one Mannheim error correcting (OMEC) codes [Hub94a, Hub94b], are briefly reviewed. The suboptimal decoding approach for Gaussian integers from [FGS13] is generalized for codes based on Eisenstein-integers. Topics treated for signal constellations based on Lipschitz and Hurwitz integers include constellation design, suboptimal detection, and set partitioning methods, and the performance on the AWGN channel is shown. A multilevel coding scheme [WFH99] for the proposed Hurwitz-integer signal constellations is discussed, where each level is protected with a non-binary low-density parity-check (LDPC) code [Gal63, DM98b].

Chapter 4 treats topics for SM with one active transmit antenna. Suboptimal active-antenna detection techniques are briefly reviewed, including signal-vector based detection (SVD) [WJS12b, PX13], the maximum-ratio combiner (MRC) [NXQ11], and the Gaussian approximation method (GAM) [LWL15]. A modification of the GAM to the modified Gaussian approximation method (MGAM) is proposed, which reduces the complexity without additional performance loss.. A comparison of the computational complexity for the detection methods is provided. Considerations for signal constellation designs are given and the detection performance is shown based on simulation results. In the remaining part of this chapter, it is shown how the simple repetition codes for transmission with SPM used in [LSL+19] can be replaced by OMEC codes. Performance gains are shown and discussed based on the simulation results.

Chapter 5 treats topics for GMSM with mutliple active transmit antennas. Suboptimal two-stage detection schemens for arbitrary multidimensional signal constellations and multidimensional signal constellations that have algebraic properties are discussed. For the latter, it is shown how the algebraic properties are exploited to reduce the detecion complexity. Detailed examples of the signal constellation design with Hurwitz integers are given. These signal sets are suitible for a GMSM transmission scheme with two active transmit antennas. A set partitioning method is shown, the computational complexity is analyzed, and the error performance for transmission with GMSM is depicted for different set sizes. The application of OMEC codes from Chapter 3 to GMSM is discussed, where a subcode is used as multidimensional signal constellation.

Chapter 6 contains a summary and concludes this thesis.

Parts of this thesis have been published in: [2–9].

Channel Models and Transmission Scenarios 2

In this chapter, the fundamental concepts of spatial modulation are introduced. We start in Chapter 2.1 with the channel models, i.e. the AWGN channel and the MIMO Rayleigh fading channel. The AWGN channel is employed in Chapter 3 to compare different signal constellations, whereas the Rayleigh fading channel is used for all MIMO transmission schemes. In Section 2.2, we discuss various spatial modulation transmission schemes from the literature, including spatial modulation, generalized spatial modulation, generalized multistream spatial modulation and the most recently-introduced spatial permutation modulation.

2.1 Channel Models

Within digital communication systems, transmitter and receiver side are connected via a communication channel. Information is transmitted over this channel through signals. The channel is a physical medium and the signals are disturbed during transmission in many ways and by various noise sources of different origins. Channel models describe the statistical properties of those disturbances. Despite the fact that communication channels are rather complex systems, the models are usually simplified and only consider some signal disturbances. All models are considered as discrete-time channels without bandwidth constraints.

2.1.1 AWGN Channel

First, we consider SISO transmission over the AWGN channel as depicted in Figure 2.1. This channel model considers additive noise only, which is the major

D. B. Rohweder, *Signal Constellations with Algebraic Properties and their Application in Spatial Modulation Transmission Schemes*, Schriftenreihe der Institute für Systemdynamik (ISD) und optische Systeme (IOS), https://doi.org/10.1007/978-3-658-37114-2_2

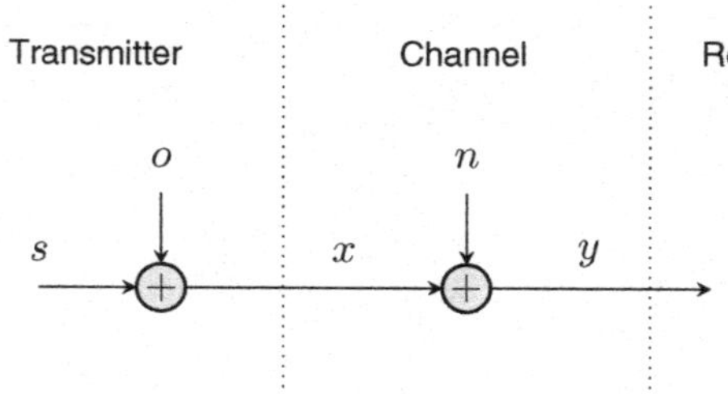

Figure 2.1 System model for the discrete-time AWGN channel including the transmitter and receiver side. The transmitter is able to apply an offset o to obtain $\mathrm{E}\{x\} = 0$

type of signal corruption for many communication channels [Pro07]. Hence, it is a perfect candidate to investigate the performance of signal constellations and detection algorithms. This thesis considers two- and four-dimensional signal constellations. For this purpose, data symbols can be taken from a two-dimensional (complex-valued) signal constellation $\mathcal{T}$, i.e. $s \in \left\{\varsigma_1, \ldots, \varsigma_{M_2}\right\} = \mathcal{T} \subset \mathbb{C}$, where M_2 denotes the cardinality of $\mathcal{T}$, or from a four-dimensional (quaternion-valued) signal constellation $\mathcal{F}$, i.e. $s \in \left\{\phi_1, \ldots, \phi_{M_4}\right\} = \mathcal{F}$, where M_4 denotes the cardinality of $\mathcal{F}$. In the following, since the given equations are valid for both cases, it is sufficient to denote and use $\mathrm{D} \in \{2, 4\}$ as the dimension. We continue with the more general case for data symbols s, taken from a D-dimensional signal constellation $\mathcal{X}$, i.e. $s \in \left\{\chi_1, \ldots, \chi_{M_\mathrm{D}}\right\} = \mathcal{X}$. If symbols are taken with equal probability, then the spectral efficiency in bits per channel use (bpcu) reads

$$\log_2(M_\mathrm{D}) \text{ bpcu} \tag{2.1}$$

and for comparison reasons, we define the data-rate unit as bits transmitted in two dimensions bits_2 as

$$\frac{2}{\mathrm{D}} \cdot \log_2(M_\mathrm{D}) \text{ bits}_2. \tag{2.2}$$

The system equation is given by

$$y = x + n, \tag{2.3}$$

where y is the received symbol, x the transmitted symbol, and n a additive white Gaussian random variable. Considering one transmission scenario, symbols are always of the same dimension. The variance of the signal constellation is given by

$$\sigma_s^2 = \mathrm{E}\left\{|s - \mathrm{E}\{s\}|^2\right\}. \tag{2.4}$$

If the signal constellation is zero-mean, i.e. $\mathrm{E}\{s\} = 0$, we can rewrite (2.4) as

$$\sigma_s^2 = \mathrm{E}\left\{|s|^2\right\}, \tag{2.5}$$

otherwise, an offset $o = -\mathrm{E}\{s\}$ is added to the symbols and thus the signal constellation is shifted to obtain zero-mean for the transmit symbols, i.e. $x = s + o$.[1] Hence, if symbols appear with equal probability over time, i.e. $\mathrm{P}_{\mathcal{X}}(\chi_1) = \mathrm{P}_{\mathcal{X}}(\chi_2) = \ldots = \mathrm{P}_{\mathcal{X}}(\chi_{M_\mathrm{D}}) = M_\mathrm{D}^{-1}$, then the variance of the transmit symbols reads

$$\sigma_x^2 = \mathrm{E}\left\{|x|^2\right\} = \frac{1}{M_\mathrm{D}} \sum_{x \in \mathcal{X}} |x|^2. \tag{2.6}$$

We define the variance of the noise term σ_n^2 per real dimension and the noise power spectral density N_0 as $N_0/2 = \sigma_n^2$. For a four-dimensional constellation, the average energy per transmit symbol E_s over N_0 acc. to [PA03, FSFF20] is given by

$$\frac{E_\mathrm{s}}{N_0} = 2\frac{\sigma_{x^2}}{\sigma_{n^2}} = \frac{\sigma_{x^2}}{2\sigma_{n^2}}, \tag{2.7}$$

where $\sigma_{n^2} = 4\sigma_{n^2}$ is the total noise variance and σ_n^2 the variance of a real-valued zero-mean white Gaussian random variable. Hence, we define the signal-to-noise ratio (SNR) as energy per information bit E_b over the noise power spectral density N_0 as

$$\frac{E_\mathrm{b}}{N_0} = \frac{E_\mathrm{s}}{N_0 \log_2(M_4)}. \tag{2.8}$$

2.1.2 Rayleigh Fading Channel

We consider the Rayleigh fading MIMO channel model, which is suitable for SM transmission schemes. We consider a communication system that employs $N_\mathrm{tx} \geq 2$ transmit antennas and $N_\mathrm{rx} \geq 2$ receive antennas. In this case, the resulting spatial channel of the system is called a MIMO channel [Pro07]. The system equation is given by

[1] In this thesis, it is assumed that signal constellations fulfill the zero-mean criterion, otherwise it is mentioned and an offset is provided.

$$\boldsymbol{y} = \sqrt{\rho}\boldsymbol{H}\boldsymbol{x} + \boldsymbol{n} \tag{2.9}$$

and more specifically by

$$\begin{bmatrix} y_1 \\ \vdots \\ y_{N_{\text{rx}}} \end{bmatrix} = \sqrt{\rho} \begin{bmatrix} h_{1,1} & \dots & h_{1,\nu} & \dots & h_{1,N_{\text{tx}}} \\ \vdots & \ddots & \vdots & & \vdots \\ h_{\tau,1} & \dots & h_{\tau,\nu} & \dots & h_{\tau,N_{\text{tx}}} \\ \vdots & & \vdots & \ddots & \vdots \\ h_{N_{\text{rx}},1} & \dots & h_{N_{\text{rx}},\nu} & \dots & h_{N_{\text{rx}},N_{\text{tx}}} \end{bmatrix} \begin{bmatrix} x_1 \\ \vdots \\ x_{N_{\text{rx}}} \end{bmatrix} + \begin{bmatrix} n_1 \\ \vdots \\ n_{N_{\text{rx}}} \end{bmatrix}. \tag{2.10}$$

The entry $h_{\tau,\nu}$ of the MIMO channel matrix $\boldsymbol{H} \in \mathbb{C}^{N_{\text{rx}} \times N_{\text{tx}}}$ corresponds to the individual channel gain from the transmit antenna ν to the receive antenna τ. We consider independent and identically-distributed (i.i.d.) Rayleigh fading channels, i.e. the entries of the channel matrix are i.i.d. complex-valued circularly-symmetric Gaussian random variables with mean $\mu_h = 0$ and variance $\sigma_h^2 = 1$. The $N_{\text{tx}} \times 1$ transmitted signal vector is denoted by $\boldsymbol{x}$ and $\boldsymbol{n}$ is the $N_{\text{rx}} \times 1$ additive white Gaussian noise vector. We use normalized transmit and noise vectors. The variance of the transmit vectors is normalized to

$$\sigma_x^2 = \mathrm{E}\left\{\rho \boldsymbol{x}^{\mathsf{H}} \boldsymbol{x}\right\} = \rho, \tag{2.11}$$

where $\boldsymbol{x}^{\mathsf{H}}$ denotes the Hermitian transpose of vector $\boldsymbol{x}$. Moreover, the entries of the noise term $\boldsymbol{n} = [n_1, \dots, n_{N_{\text{rx}}}]^{\mathsf{T}}$ are i.i.d. complex-valued circularly-symmetric Gaussian random variables with mean $\mu_n = 0$ and variance $\sigma_n^2 = 1$. Hence, the pseudo SNR reads

$$\frac{\sigma_x^2}{\sigma_n^2} = \rho. \tag{2.12}$$

2.2 The Fundamental Concept of Spatial Modulation

Spatial modulation (Figure 2.2) is a MIMO transmission technique that uses only a single active transmit antenna in each time slot [MHS+08, JGS08, YDX+15]. With generalized spatial modulation and generalized multistream spatial modulation, a subset of the transmit antennas is used for each transmitted symbol [YSMH10, NDPNV13, CSSS15]. In order to improve the spectral efficiency of SM, the selection of the subset is used to transmit information.

The main difference between the various SM transmission schemes and conventional MIMO transmission is that with SM there are some zero elements in the transmit-signal vector $\boldsymbol{x}$, i.e. not all transmit antennas are activated simultaneously. Information is transmitted by choosing symbols from a signal constellation and additionally choosing a subset of transmit antennas. With GSM and GMSM, at least two transmit antennas are active, simultaneously transmitting the same or independent data symbols over each active transmit antenna.

At the receiver, ML detection or a suboptimal detection can be used to retrieve the transmitted symbol vector. Detection algorithms are discussed in Section 4.1.1 for SM and Section 5.1 for GSM and GMSM. The spatial modulation transmission schemes are subsequently introduced.

2.2.1 Spatial Modulation

With SM, only one transmit antenna is active per time instance [MHAY06, MHS+08]. Hence, the transmitted signal vector

$$\boldsymbol{x} = [x_1, x_2, \ldots, x_\nu, \ldots, x_{N_{\text{tx}}}]^{\mathsf{T}} = [\ldots, 0, s, 0, \ldots]^{\mathsf{T}}, \text{ with } x_\nu = s \in \mathcal{T}, \quad (2.13)$$

contains a single element s from a complex-valued M_2-ary signal constellation $\mathcal{T} \subset \mathbb{C} \setminus \{0\}$. Note that the zero element is used as an indication of inactive antennas and thus cannot be used as an element in $\mathcal{T}$. Due to (2.13), we can write

$$\boldsymbol{y} = \sqrt{\rho}\boldsymbol{h}_i s + \boldsymbol{n}, \; i \in \mathcal{A}, \quad (2.14)$$

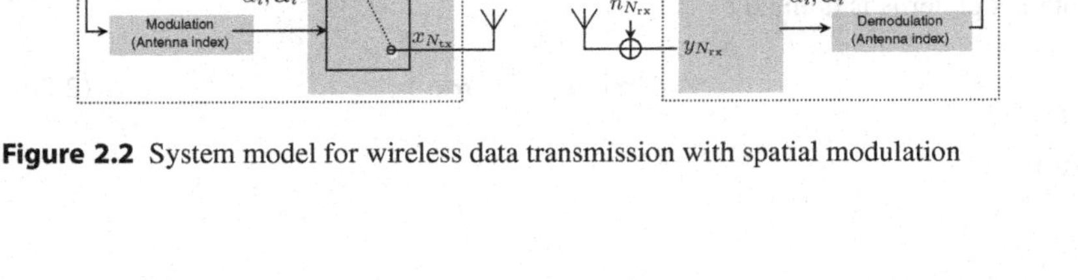

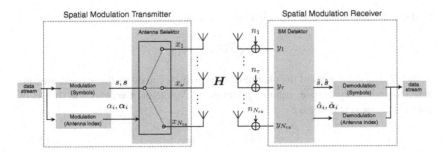

Figure 2.2 System model for wireless data transmission with spatial modulation

where $\boldsymbol{h}_i$ is the ith column of $\boldsymbol{H}$. Note that the index i corresponds to the so-called antenna pattern α_i, which defines the active transmit antenna, cf. (2.10). The set of antenna patterns $\mathcal{A}$ is defined as

$$\mathcal{A} = \left\{\alpha_1, \alpha_2, \ldots, \alpha_{N_\mathrm{p}}\right\} \subseteq \{1, 2, \ldots, N_\mathrm{tx}\} \subset \mathbb{N}, \tag{2.15}$$

with the cardinality or total number of applied antenna patterns

$$1 \leq N_\mathrm{p} = |\mathcal{A}| \leq N_\mathrm{tx}. \tag{2.16}$$

Information is transmitted by selecting the antenna pattern $\alpha_i \in \mathcal{A}$ as well as selecting one symbol $s \in \mathcal{T}$. The spectral efficiency reads

$$\log_2(N_\mathrm{p} M_2) \text{ bpcu}. \tag{2.17}$$

2.2.2 Generalized Spatial Modulation

If we consider the transmission of binary data streams, it is obvious that SM has the disadvantage that the number of transmit antennas has to be a power of two; otherwise, spectral efficiency remains unexploited. GSM [YSMH10] gives more flexibility in the design of antenna patterns for a given multiple-antenna setup. With GSM, two or more transmit antennas are active, sending the same symbol per time instance. Hence, the transmitted signal vector

$$\boldsymbol{x} = [\ldots, 0, \ldots, s, \ldots, s, \ldots, 0, \ldots]^\mathsf{T}, \text{ with } s \in \mathcal{T} \subset \mathbb{C} \setminus \{0\}, \tag{2.18}$$

contains N_a-times the non-zero element s. In analogy to (2.14), we can write

$$\boldsymbol{y} = \sqrt{\rho}\boldsymbol{H}_i s + \boldsymbol{n}, \tag{2.19}$$

where $\boldsymbol{H}_i$ is a submatrix of the channel matrix, which considers the active transmit-antenna paths only. Hence, antenna patterns are N_a-element vectors and the set of antenna patterns is defined as

$$\mathcal{A} = \left\{\boldsymbol{\alpha}_1, \boldsymbol{\alpha}_2, \ldots, \boldsymbol{\alpha}_{N_\mathrm{p}}\right\}, \tag{2.20}$$

with

$$\boldsymbol{\alpha}_i \in \left\{\left[\alpha_{i,1}, \alpha_{i,2}, \ldots, \alpha_{i,N_\mathrm{a}}\right] \middle| 1 \leq \alpha_{i,1} < \alpha_{i,2} < \ldots < \alpha_{i,N_\mathrm{a}} \leq N_\mathrm{tx}\right\} \subseteq \mathbb{N}^{1\times N_\mathrm{a}}. \tag{2.21}$$

The cardinality or total number of applied antenna patterns is given by the inequality

$$1 \leq N_\mathrm{p} = |\mathcal{A}| \leq \binom{N_\mathrm{tx}}{N_\mathrm{a}}. \tag{2.22}$$

Typically, only a subset of these possibilities is used to map $N_\mathrm{p} = \lfloor \log_2 \binom{N_\mathrm{tx}}{N_\mathrm{a}} \rfloor$ bpcu onto the spatial data component. Information is transmitted by selecting the antenna pattern $\boldsymbol{\alpha}_i \in \mathcal{A}$ as well as selecting one symbol $s \in \mathcal{T}$. As for SM, the spectral efficiency is given by (2.17).

2.2.3 Generalized Multistream Spatial Modulation

GMSM [WJS12a] is a transmission technique similar to GSM. Comparing the two, with GMSM symbols are chosen independently for each active transmit antenna, which results in a higher spectral efficiency per channel use for the same multiple-antenna setup. Furthermore, GSM can be considered as a coded-GMSM transmission scheme, where the symbol vector represents codewords of a repetition code. With GMSM, the transmitted signal vector

$$\boldsymbol{x} = [\ldots, 0, \ldots, s_1, \ldots, 0, \ldots, s_{N_\mathrm{a}}, \ldots, 0, \ldots]^\mathsf{T}, \tag{2.23}$$

contains the complex-valued N_a-element symbol vector

$$\boldsymbol{s} \in \mathcal{X} = \left\{[s_1, s_2, \ldots, s_{N_\mathrm{a}}]^\mathsf{T} \middle| s_j \neq 0, j = 1, 2, \ldots, N_\mathrm{a}\right\} \subset \mathbb{C}^{N_\mathrm{a}\times 1}. \tag{2.24}$$

Throughout this thesis, various construction methods to design signal constellations are presented. Some of them are obtained from lattices—later introduced—or a set of points defined in the real D-dimensional space $\mathbb{R}^\mathrm{D}$. However, given that the considered antennas take complex-valued symbols, it is convenient to define a common multidimensional M_D-ary signal constellation as set of points $\mathfrak{x}$ as

$$\mathfrak{X} = \left\{\left[\mathfrak{x}_1, \mathfrak{x}_2, \ldots, \mathfrak{x}_{2N_\mathrm{a}}\right]^\mathsf{T} \middle| |\mathfrak{x}_{k-1}| + |\mathfrak{x}_k| \neq 0 \, \forall k \in \{2j \mid j \in \mathbb{N}, j \leq 2N_\mathrm{a}\}\right\} \subset \mathbb{R}^{2N_\mathrm{a}\times 1} \tag{2.25}$$

in the real D-dimensional space of dimension $\mathrm{D} = 2N_\mathrm{a}$. The mapping $\mathfrak{x} \rightarrow \boldsymbol{s}$ between points in real space and the complex-valued symbol vector is defined as

$$s = [s_1, s_2, \ldots, s_{N_a}]^T \equiv [\mathfrak{x}_1 + i\mathfrak{x}_2, \mathfrak{x}_3 + i\mathfrak{x}_4, \ldots, \mathfrak{x}_{2N_a-1} + i\mathfrak{x}_{2N_a}]^T. \quad (2.26)$$

Hence, we have defined the mapping $\mathfrak{X} \to \mathcal{X}$.

The system equation and the set of antenna patterns follow the same procedure as already introduced for GSM in (2.19)–(2.22). Information is transmitted by selecting the antenna pattern $\boldsymbol{\alpha}_i \in \mathcal{A}$ as well as selecting one symbol vector $s \in \mathcal{X}$. The spectral efficiency depends on the cardinality of the set of antenna patterns N_p as well as the number of symbol vectors in $\mathcal{X}$, and it is given by

$$\log_2(N_p M_D) \text{ bpcu}. \quad (2.27)$$

Note that it is common practice to utilize one complex-valued signal constellation, e.g. M_2-QAM or M_2-PSK, to create the symbol vector. Then, the symbol vector is given as

$$s = [s_1, s_2, \ldots, s_{N_a}]^T, \text{ with } s_{1,2,\ldots,N_a} \in \mathcal{T} \subset \mathbb{C} \setminus \{0\} \quad (2.28)$$

and the spectral efficiency reads

$$\log_2(N_p M_2^{N_a}) \text{ bpcu}. \quad (2.29)$$

Figure 2.3 depicts the greatest possible spectral efficiency depending on the number of transmit antennas for the transmission with SM, GSM, and GMSM. The signal

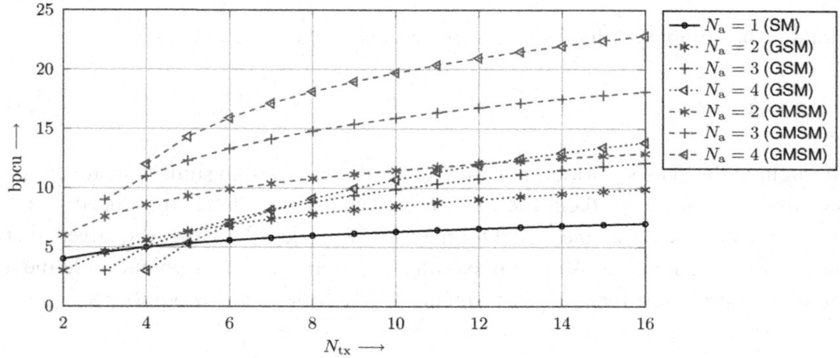

Figure 2.3 The maximum spectral efficiency in bpcu as a function of number of transmit antennas N_{tx} for SM (solid line), GSM (dotted lines) and GMSM (dashed lines). The signal constellation contains M_2=8 symbols

constellation contains M_2=8 symbols. Note that since we apply a complex-valued (two-dimensional) signal constellation for all transmission schemes, the symbol vector for the GMSM transmission scheme is given by (2.28). As can be seen from the figure, for SM and GSM when N_a is close to N_{tx}, the number of possible antenna patterns is small and hence the contribution to the overall spectral efficiency is small. Thus, for some scenarios it is possible to increase the spectral efficiency by reducing the number of active transmit antennas.

2.2.4 Spatial Permutation Modulation

SPM is a recently proposed transmission technique based on SM [LSL+19]. In contrast to SM, the spatial symbol is dispersed in time, i.e. the channel is used T-times to transmit one information symbol. As with SM, only one antenna is active per channel use, and hence $T = 1$ corresponds to the conventional SM transmission scheme. The so-called transmission pattern

$$\boldsymbol{p}_i = \left[p_{i,1}, p_{i,2}, \ldots, p_{i,T}\right]^{\mathrm{T}}, \text{ with } p_{i,\iota} \in \{1, 2, \ldots, N_{\mathrm{tx}}\} \tag{2.30}$$

indicates the active transmit antenna at successive time instances. Transmit antennas are active at most once during a sequence. In analogy to the other spatial modulation transmission schemes, we use N_p to denote the number of transmission patterns applied. The set of transmission patterns is given by

$$\mathcal{P} = \left\{\boldsymbol{p}_1, \boldsymbol{p}_2, \ldots, \boldsymbol{p}_{N_\mathrm{p}}\right\}. \tag{2.31}$$

The largest possible set $\mathcal{P}$ includes all possible ordered arrangements of T elements from an N_{tx}-element vector with $1 \leq T \leq N_{tx}$, and hence the largest set includes

$$N_{\mathrm{p}_{\max}} = \frac{N_{\mathrm{tx}}!}{(N_{\mathrm{tx}} - T)!} \tag{2.32}$$

transmission patterns. Besides other properties, the error performance with this transmission scheme depends on the minimum Hamming distance dist_H for a given set. It is shown in [LSL+19] that a higher minimum Hamming distance may lead to a lower error probability. The minimum Hamming distance for a given set reads [Bos99]

$$\mathrm{dist_H} = \min_{\substack{\mathbf{p}_j,\mathbf{p}_k \in \mathcal{P} \\ j \neq k}} \sum_{\iota=1}^{T} \mathrm{wt}\left(p_{j,\iota} + p_{k,\iota}\right), \text{ with } \mathrm{wt}\left(p_{j,\iota} + p_{k,\iota}\right) = \begin{cases} 0, & \text{for } p_{j,\iota} = p_{k,\iota} \\ 1, & \text{for } p_{j,\iota} \neq p_{k,\iota} \end{cases}. \tag{2.33}$$

For $T = N_{\mathrm{tx}}$ and $T = N_{\mathrm{tx}} - 1$, we have $N_{\mathrm{p_{max}}} = N_{\mathrm{tx}}!$ transmission patterns within both sets with a minimum Hamming distance of $\mathrm{dist_H} = 2$ and $\mathrm{dist_H} = 1$, respectively. This is due to the fact that the first mentioned can take one more element for every transmission pattern, and even if the cardinality of the set cannot be increased, it increases the minimum Euclidean distance. Example 2.1 provides the construction of two sets of transmission patterns and addresses the aforementioned considerations.

Example 2.1 *Let $N_{\mathrm{tx}} = 3$ and $T = 2$ for the first set $\mathcal{P}_1$ and $T = 3$ for the second set $\mathcal{P}_2$. The largest sets include the following $N_{\mathrm{p_{max}}} = 3! = 6$ transmission patterns:*

$$\mathcal{P}_1 = \left\{ \begin{bmatrix} 1 \\ 2 \end{bmatrix}, \begin{bmatrix} 2 \\ 1 \end{bmatrix}, \begin{bmatrix} 1 \\ 3 \end{bmatrix}, \begin{bmatrix} 3 \\ 1 \end{bmatrix}, \begin{bmatrix} 2 \\ 3 \end{bmatrix}, \begin{bmatrix} 3 \\ 2 \end{bmatrix} \right\}, \text{ with } \mathrm{dist_H} = 1$$

and

$$\mathcal{P}_2 = \left\{ \begin{bmatrix} 1 \\ 2 \\ 3 \end{bmatrix}, \begin{bmatrix} 2 \\ 1 \\ 3 \end{bmatrix}, \begin{bmatrix} 1 \\ 3 \\ 2 \end{bmatrix}, \begin{bmatrix} 3 \\ 1 \\ 2 \end{bmatrix}, \begin{bmatrix} 3 \\ 2 \\ 1 \end{bmatrix}, \begin{bmatrix} 2 \\ 3 \\ 1 \end{bmatrix} \right\}, \text{ with } \mathrm{dist_H} = 2.$$

For $T = N_{\mathrm{tx}}$, the minimum Hamming distance is always $\mathrm{dist_H} \geq 2$.

Note that reducing the number of transmission patterns in the set can improve the minimum Hamming distance.

Table 2.1 provides sets of transmission patterns for various T, N_{p}, and $\mathrm{dist_H}$. Such sets are used in [LSL+19] to construct space-time block codes. Most of the sets are given in [LSL+19], but some additional sets were found by computer search. Note that different sets of transmission patterns with the same parameters may exist. The selected sets have an equilibrium in average antenna usage to achieve a high transmit diversity.

With SPM, the SM system model is extended by time components. The $N_{\mathrm{tx}} \times T$ signal matrix

$$\boldsymbol{X}\left(\boldsymbol{p}_i, \boldsymbol{s}\right) = [\boldsymbol{x}_1, \boldsymbol{x}_2, \ldots, \boldsymbol{x}_\iota, \ldots, \boldsymbol{x}_T] = \begin{bmatrix} x_{1,1} & \cdots & x_{1,\iota} & \cdots & x_{1,T} \\ \vdots & \ddots & \vdots & & \vdots \\ x_{\nu,1} & \cdots & x_{\nu,\iota} & \cdots & x_{\nu,T} \\ \vdots & & \vdots & \ddots & \vdots \\ x_{N_{\mathrm{tx}},1} & \cdots & x_{N_{\mathrm{tx}},\iota} & \cdots & x_{N_{\mathrm{tx}},T} \end{bmatrix} \quad (2.34)$$

is constructed by combining a transmission pattern $\boldsymbol{p}_i \in \mathcal{P}$ with a symbol vector

$$\boldsymbol{s} = [s_1, s_2, \ldots, s_\iota, \ldots, s_T], \text{ with } s_\iota \in \mathbb{C} \setminus \{0\}, \quad (2.35)$$

where ι corresponds to a specific time instance with $1 \leq \iota \leq T$. At this point, we can construct the symbol vector comprising independent symbols from a signal constellation $s_\iota \in \mathcal{T}$ or use a codeword, i.e. $\boldsymbol{s} \in \mathcal{C}$. Hence, the corresponding linear block code is of length T, dimension k, and rate $R = k/T$. The signal matrix contains

Table 2.1 Sets of transmission patterns for various T, N_{p} and $\mathrm{dist}_{\mathrm{H}}$ for a multiple-antenna setup with $N_{\mathrm{tx}} = 4$ transmit antennas

T	N_{p}	$\mathrm{dist}_{\mathrm{H}}$	Set of transmission pattern $\mathcal{P}$
3	1	–	$[1,2,3]^{\mathsf{T}}$
	4	3	$[1,2,3]^{\mathsf{T}}, [2,3,4]^{\mathsf{T}}, [3,4,1]^{\mathsf{T}}, [4,1,2]^{\mathsf{T}}$
	8	2	$[1,2,3]^{\mathsf{T}}, [1,3,4]^{\mathsf{T}}, [2,1,4]^{\mathsf{T}}, [3,4,1]^{\mathsf{T}}$, $[2,4,3]^{\mathsf{T}}, [3,1,2]^{\mathsf{T}}, [4,2,1]^{\mathsf{T}}, [4,3,2]^{\mathsf{T}}$
	16	1	$[1,2,3]^{\mathsf{T}}, [1,3,2]^{\mathsf{T}}, [1,4,2]^{\mathsf{T}}, [1,4,3]^{\mathsf{T}}$, $[2,1,3]^{\mathsf{T}}, [2,1,4]^{\mathsf{T}}, [2,3,4]^{\mathsf{T}}, [2,4,1]^{\mathsf{T}}$, $[3,1,4]^{\mathsf{T}}, [3,2,1]^{\mathsf{T}}, [3,2,4]^{\mathsf{T}}, [3,4,1]^{\mathsf{T}}$, $[4,1,2]^{\mathsf{T}}, [4,2,3]^{\mathsf{T}}, [4,3,1]^{\mathsf{T}}, [4,3,2]^{\mathsf{T}}$
	24	1	$[1,2,3]^{\mathsf{T}}, [1,3,2]^{\mathsf{T}}, \ldots$ (all possible arrangements)
4	1	–	$[1,2,3,4]^{\mathsf{T}}$
	4	4	$[1,2,3,4]^{\mathsf{T}}, [2,3,4,1]^{\mathsf{T}}, [3,4,1,2]^{\mathsf{T}}, [4,1,2,3]^{\mathsf{T}}$
	8	3	$[1,2,3,4]^{\mathsf{T}}, [1,3,4,2]^{\mathsf{T}}, [1,4,2,3]^{\mathsf{T}}, [2,3,1,4]^{\mathsf{T}}$, $[2,4,3,1]^{\mathsf{T}}, [2,1,4,3]^{\mathsf{T}}, [3,2,4,1]^{\mathsf{T}}, [4,1,3,2]^{\mathsf{T}}$
	16	2	$[1,2,3,4]^{\mathsf{T}}, [1,2,4,3]^{\mathsf{T}}, [1,3,2,4]^{\mathsf{T}}, [1,3,4,2]^{\mathsf{T}}$, $[1,4,2,3]^{\mathsf{T}}, [1,4,3,2]^{\mathsf{T}}, [2,1,3,4]^{\mathsf{T}}, [2,1,4,3]^{\mathsf{T}}$, $[2,3,1,4]^{\mathsf{T}}, [2,3,4,1]^{\mathsf{T}}, [3,1,2,4]^{\mathsf{T}}, [3,1,4,2]^{\mathsf{T}}$, $[3,2,1,4]^{\mathsf{T}}, [3,2,4,1]^{\mathsf{T}}, [3,4,1,2]^{\mathsf{T}}, [3,4,2,1]^{\mathsf{T}}$
	24	2	$[1,2,3,4]^{\mathsf{T}}, [1,2,4,3]^{\mathsf{T}}, \ldots$ (all possible arrangements)

one symbol s_ι in every column $\boldsymbol{x}_\iota$. The vertical position of s_ι in $\boldsymbol{x}_\iota$ is determined by the vector $\boldsymbol{p}_i$ and it defines the designated transmit antenna ν for transmission.

Example 2.2 *We demonstrate the construction of the signal matrix* $\boldsymbol{X}$ *for the setup* $N_{\mathrm{tx}} = 4$, $T = 3$. *For these parameters, there are five possible sets of transmission patterns in Table 2.1. For instance, consider the pattern* $[3, 4, 1]^{\mathrm{T}}$ *for* $N_{\mathrm{p}} = 4$. *This pattern results in the signal matrix*

$$\boldsymbol{X}(\boldsymbol{p}_3, \boldsymbol{s}) = \begin{bmatrix} 0 & 0 & s_3 \\ 0 & 0 & 0 \\ s_1 & 0 & 0 \\ 0 & s_2 & 0 \end{bmatrix}, \tag{2.36}$$

where s_1, s_2, s_3 *are symbols from the complex-valued signal constellation* $\mathcal{T}$. *These symbols can be symbols of a short block code, e.g. a repetition code. In the case of a repetition code, we have* $s_1 = s_2 = s_3$.

For the SPM system model, the noise term is extended to the noise matrix

$$\boldsymbol{N} = \begin{bmatrix} n_{1,1} & \cdots & n_{1,T} \\ \vdots & \ddots & \vdots \\ n_{N_{\mathrm{rx}},1} & \cdots & n_{N_{\mathrm{rx}},T} \end{bmatrix}. \tag{2.37}$$

The entries of the $N_{\mathrm{rx}} \times T$ noise matrix and the $N_{\mathrm{rx}} \times N_{\mathrm{tx}}$ channel matrix $\boldsymbol{H}$ have the same properties as for SM. The $N_{\mathrm{rx}} \times T$ received signal matrix $\boldsymbol{Y}$ at the receiver is given as

$$\boldsymbol{Y} = \sqrt{\rho}\boldsymbol{H}\boldsymbol{X} + \boldsymbol{N}. \tag{2.38}$$

Information is transmitted by selecting the transmission pattern $\boldsymbol{p}_i \in \mathcal{P}$, as well as selecting the symbols s_ι for the symbol vector $\boldsymbol{s}$. Hence, the maximum spectral efficiency in bits per sequence (bpsq) for the code rate $R = 1$ and a given signal constellation with M_2 elements reads

$$\log_2(\sqrt[T]{N_{\mathrm{p}}}M_2)\ \mathrm{bpcu} = \log_2(N_{\mathrm{p}}M_2^T)\ \mathrm{bpsq}, \tag{2.39}$$

where we can clearly see the relation 1 bpcu $= T$ bpsq.

Signal Constellations with Algebraic Properties

3

In this chapter, we present new code-construction methods, suboptimal detection and set-partitioning techniques and analyze the error performance on the AWGN channel. The application of signal constellations with algebraic properties to SM is treated in Chapter 4 for two-dimensional and Chapter 5 for multidimensional signal constellations.

In the first section, we briefly summarize codes over finite Gaussian and Eisenstein integer fields, first introduced by Huber [Hub94a, Hub94b]. Gaussian and Eisenstein integers are subsets of the complex numbers. Hence, complex-valued finite fields or rings are directly applicable as signal constellations in communication systems, instead of using conventional signal constellations, e.g. square QAM or PSK. Subsequently, we generalize a suboptimal detection approach for Eisenstein integers, which was originally proposed for Gaussian integers in [FGS13]. In Section 5.3, we revise this approach for GMSM based transmission schemes. Moreover, we discuss the construction of LDPC codes over finite Gaussian integer fields.

In the second section, four-dimensional signal constellations based on Lipschitz and Hurwitz integers are discussed. Lipschitz and Hurwitz integers are subsets of the quaternions, which can be considered as an extension of the complex numbers with one real and three imaginary parts. Such signal constellations can be applied in optical communications and with dual-polarized antennas in wireless communication systems, where the transmission is based on the in-phase and quadrature components of both polarizations [KA10, SFFF19, FSFF20]. This corresponds to the independent transmission of two complex-valued symbols. It is shown that four-dimensional signal constellations even with suboptimal detection at the receiver can outperform two-dimensional signal constellations with ML detection in terms of the symbol error rate (SER) in the high-SNR range.

D. B. Rohweder, *Signal Constellations with Algebraic Properties and their Application in Spatial Modulation Transmission Schemes*, Schriftenreihe der Institute für Systemdynamik (ISD) und optische Systeme (IOS), https://doi.org/10.1007/978-3-658-37114-2_3

At the end of this chapter, two set-partitioning techniques are discussed that pursue different aims. One reduces the computational complexity for the signal detection procedure, whereas the other is used to find subsets with high minimum squared Euclidean distances. The latter mentioned is used in a non-binary multilevel coding scheme. An example with two levels is presented in Section 3.2.6, where simulation results show that a significant coding gain can be achieved.

Parts of this chapter have already been published in: [4–6, 8].

3.1 Two-Dimensional Signal Constellations

In the following subsections, we first introduce Gaussian and Eisenstein integers and some corresponding code constructions. A suboptimal decoding procedure is provided, where simulation results show that near-ML performance can be achived.

3.1.1 Codes over Gaussian and Eisenstein Integers

Gaussian integers are complex numbers where the real and imaginary parts are integers, i.e. the set of Gaussian integers is

$$\mathcal{G} = \{a + b\mathrm{i} \mid a, b \in \mathbb{Z}\} \subset \mathbb{C}, \tag{3.1}$$

with $\mathrm{i}^2 = -1$. They are isomorphic to the two-dimensional integer lattice [CS99]. The norm of a Gaussian integer z is given as

$$N(z) = zz^* = a^2 + b^2, \tag{3.2}$$

where $z^* = a - b\mathrm{i}$ is the complex conjugate of z. A non-zero Gaussian integer $\pi \in \mathcal{G} \setminus \{0\}$ has the inverse

$$\pi^{-1} = \frac{\pi^*}{N(\pi)}. \tag{3.3}$$

Hence, we are able to define a modulo function as [Hub94b]

$$z \bmod \pi = z - \lfloor z\pi^{-1} \rceil \pi, \tag{3.4}$$

where $\lfloor \cdot \rceil$ denotes rounding to the closest integer, i.e. for a real number $a \in \mathbb{R}$, $\lfloor a \rceil \in \mathbb{Z}$ yields the integer closest to it, where we round half-way values towards plus infinity. For a complex number $z = a + b\mathrm{i}$, we use the rounding $\lfloor z \rceil = \lfloor a \rceil + \lfloor b \rceil \mathrm{i}$.

Most known code constructions for Gaussian integers are linear codes based on finite Gaussian-integer fields $\mathcal{G}_p$, which are constructed from primes p such that $p - 1$ is divisible by four [Hub94b]. These primes are the sum of two perfect squares, i.e. $p = a^2 + b^2$ with $a, b \in \mathbb{Z}$. Hence, they can be represented as the product $p = \pi\pi^*$ with $\pi = a + bi$. The finite Gaussian-integer field is the set

$$\mathcal{G}_p = \{z \bmod \pi \mid z = 0, \dots, p-1, z \in \mathbb{Z}\} \subset \mathscr{G}. \tag{3.5}$$

Similarly, the set of all Eisenstein integers $\mathscr{E}$ is given as

$$\mathscr{E} = \{a + b\omega |\, a, b \in \mathbb{Z}\} \subset \mathbb{C}, \tag{3.6}$$

where the basis element $\omega = \frac{1}{2}(-1 + \sqrt{3}\mathrm{i}) = \mathrm{e}^{\mathrm{i}2\pi/3}$ is a complex root of unity. With the complex conjugate $\omega^* = \omega^2 = \frac{1}{2}(-1 - \sqrt{3}\mathrm{i})$, we have $|\omega|^2 = \omega\omega^* = 1$. The Eisenstein integers form the triangular or hexagonal lattice in the complex plane [CS99], which is the densest packing in two dimensions. Eisenstein integers have the norm value $N(z) = |z|^2 = zz^*$ with the complex conjugate $z^* = a + b\omega^*$. Consequently, with $\lambda^{-1} = \frac{\lambda^*}{N(\lambda)}$, we can define a modulo function

$$z \bmod \lambda = z - \left[z\lambda^{-1}\right]\lambda\,, \tag{3.7}$$

where $[\cdot]$ denotes rounding to the closest Eisenstein integer such that the norm of this integer is as small as possible [Hub94a].

Note that primes p, such that $p - 1$ is divisible by six, can be represented as the product of two conjugate Eisenstein integers $p = \lambda\lambda^* = N(\lambda)$ [Hub94a]. For such primes p, we are able to define finite fields over Eisenstein integers

$$\mathcal{E}_p = \{z \bmod \lambda \mid z = 0, \dots, p-1, z \in \mathbb{Z}\} \subset \mathscr{E}. \tag{3.8}$$

Example 3.1 *We provide two examples of these constellations, which will later on be used to construct codes. The prime* $p = 13$ *satisfies the condition* $13 \bmod 4 = 1$. *It can be factorized as* $13 = (3+2\mathrm{i})(3-2\mathrm{i}) = 3^2+2^2$. *Hence, with* $\pi = 3 + 2\mathrm{i}$ *we will construct the set* $\mathcal{G}_{13}$ *of 13 Gaussian integers that form a finite field with respect to complex-valued addition and multiplication modulo* π.

Similarly, the prime $p = 19$ satisfies the condition $19 \bmod 6 = 1$. This prime can be factorized as $19 = (5+2\omega)(5+2\omega^2)$. We will use this factorization to construct the finite field $\mathcal{E}_{19}$ of 19 Eisenstein integers with $\lambda = 5+2\omega$. Both constellations are depicted in Figure 3.1 drawn in the complex plane.

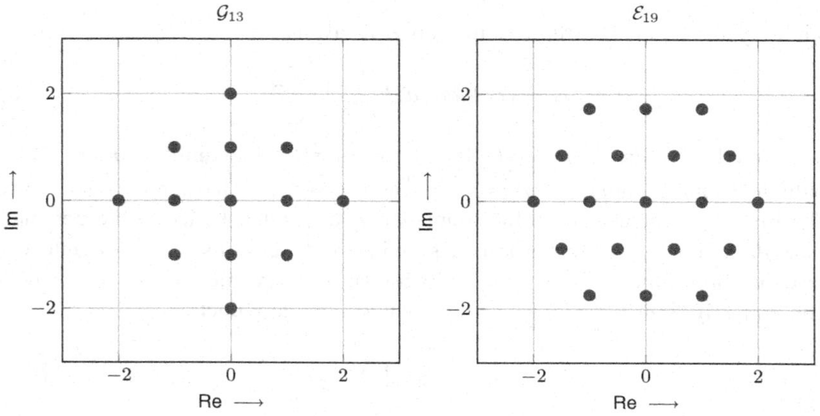

Figure 3.1 Gaussian-integer constellation $\mathcal{G}_{13}$ (left) and Eisenstein-integer constellation $\mathcal{E}_{19}$ (right) drawn in the complex plane

The class $\mathcal{K}(z)$ of a Gaussian or Eisenstein integer z is the set of all numbers z' such that $z = z' \bmod \pi$ or $z = z' \bmod \lambda$, respectively. The Mannheim weight of a Gaussian integer z is defined as [FGS13, MBG07]

$$\mathrm{wt_M}(z) = \min_{a+bi \in \mathcal{K}(z)} |a| + |b| \tag{3.9}$$

and the Mannheim weight of a vector $\boldsymbol{v} = \left[v_0, v_1, \ldots, v_{t-1}\right]$ is

$$\mathrm{wt_M}(\boldsymbol{v}) = \sum_{j=0}^{t-1} \mathrm{wt_M}(v_j). \tag{3.10}$$

Similarly, we use the weight

$$\mathrm{wt_M}(z) = \min_{a+b\omega \in \mathcal{K}(z)} |a| + |b| \tag{3.11}$$

for Eisenstein integers. With this weight function, we can define the distance between two Gaussian or Eisenstein integers v and z as

$$\mathrm{dist_M}(v, z) = \mathrm{wt_M}(v - z) \tag{3.12}$$

and the distance between the vectors $\boldsymbol{v} = [v_0, v_1, ..., v_{t-1}]$ and $\boldsymbol{z} = [z_0, z_1, ..., z_{t-1}]$ is

$$\mathrm{dist_M}(\boldsymbol{v}, \boldsymbol{z}) = \sum_{j=0}^{t-1} \mathrm{dist_M}(v_j, z_j) = \mathrm{wt_M}(\boldsymbol{v} - \boldsymbol{z}). \tag{3.13}$$

Let α be a primitive element of a Gaussian-integer or Eisenstein-integer field. A one-Mannheim error-correcting (OMEC) code of length N_c over such fields is defined by its parity-check matrix [Hub94b, Hub94a]

$$\boldsymbol{P} = \left[\alpha^0, \alpha^1, \ldots, \alpha^{N_c-1}\right], \tag{3.14}$$

where the elements are generated by powers of α. This code has dimension $K = N_c - 1$. All codewords $\boldsymbol{c} = [c_0, c_1, \ldots, c_{N_c-1}]$ satisfy

$$\boldsymbol{P}\boldsymbol{c}^\mathsf{T} = 0. \tag{3.15}$$

For codes over Gaussian integers, we have $c_j \in \mathcal{G}_p$ and the maximum code length is

$$N_c = \frac{p-1}{4}. \tag{3.16}$$

Similarly, for Eisenstein integers, we have $c_j \in \mathcal{E}_p$ and the maximum length

$$N_c = \frac{p-1}{6}. \tag{3.17}$$

Systematic encoding with the information vector $\boldsymbol{u} = [u_0, \ldots, u_{K-1}]$ is obtained by

$$\begin{aligned} \boldsymbol{c} &= [c_0, u_0, u_1 \ldots, u_{K-1}] \text{ with} \\ c_0 &= -\alpha^1 u_0 - \alpha^2 u_1 - \ldots - \alpha^{N_c-1} u_{K-1}. \end{aligned} \tag{3.18}$$

An OMEC code $\mathcal{C}(N_c, K)$ has a minimum Hamming distance $\text{dist}_\text{H} = 2$ and minimum Mannheim distance $\text{dist}_\text{M} = 3$, and hence such a code can detect any single error with arbitrary Mannheim weight and correct a single error of Mannheim weight one. Note that this includes all errors where a symbol is mapped by the channel to a nearest neighbor. OMEC codes of maximum length attain the sphere-packing bound for the Mannheim distance [Hub94b, Hub94a].

3.1.2 Decoding of OMEC Codes for the AWGN Channel

First we consider the ML-decoding method, which always gives the optimal decision for the most likely sent codeword if data symbols are chosen uniformely distributed at the transmitter. We consider transmission over the AWGN channel as described in Section 2.1.1, where

$$\boldsymbol{y} = [y_0, y_1, \ldots, y_{N_c-1}] \tag{3.19}$$

denotes the received vector, which includes N_c transmissions. For a given OMEC code $\mathcal{C}$, the ML-decoding rule at the receiver reads

$$\hat{\boldsymbol{c}} = \underset{\boldsymbol{c} \in \mathcal{C}}{\operatorname{argmin}} \, ||\boldsymbol{y} - \boldsymbol{c}||^2. \tag{3.20}$$

It is associated with the highest computational complexity compared with other decoding methods. Suboptimal decoding methods may be necessary, e.g. for high spectral efficiency or low-latency applications.

Next, we introduce a list-decoding algorithm for suboptimal soft-input decoding of OMEC codes. This method was proposed in [FGS13] for OMEC codes over Gaussian-integer fields and transmission over the AWGN channel. However, in the following we will show that this decoding method can also be applied to OMEC codes over Eisenstein-integer fields. In Section 5.3, we generalize the decoding method for GMSM transmission.

Let

$$\tilde{\boldsymbol{y}} = \lfloor \boldsymbol{y} \rceil, \tilde{y}_j \in \mathcal{G}_p \tag{3.21}$$

or

$$\tilde{\boldsymbol{y}} = [\boldsymbol{y}], \tilde{y}_j \in \mathcal{E}_p \tag{3.22}$$

be the hard-input vector at the receiver, i.e. the elements of $\tilde{\boldsymbol{y}}$ are mapped to the closest symbol within the Gaussian-integer or Eisenstein-integer constellation with respect to the squared Euclidean distance given in (3.39). Note that OMEC codes of maximum code length are perfect with respect to the Mannheim distance. Hence, syn-

Table 3.1 Syndrome lookup table for the Gaussian code from Example 3.2

syndrome s	1	2	$1+\mathrm{i}$
error vector $\boldsymbol{e}$	$[-1, 0, -\mathrm{i}]$	$[2, 0, 0]$	$[1+\mathrm{i}, 0, 0]$
	$[0, 0, -1+\mathrm{i}]$	$[-1, 0, -1]$	$[0, \mathrm{i}, -\mathrm{i}]$
	$[0, \mathrm{i}, 1]$	$[0, \mathrm{i}, \mathrm{i}]$	$[0, 0, -2]$
	$[\mathrm{i}, -\mathrm{i}, 0]$	$[\mathrm{i}, 0, 1]$	$[-1, -1, 0]$
	$[-\mathrm{i}, 1, 0]$	$[0, 1-\mathrm{i}, 0]$	$[-\mathrm{i}, 0, \mathrm{i}]$
	$[0, -2, 0]$	$[-\mathrm{i}, -1, 0]$	$[0, -\mathrm{i}, 1]$

drome decoding based on the Mannheim distance is optimal for hard-input decoding. The syndrome

$$s = \boldsymbol{P}\tilde{\boldsymbol{y}}^{\mathsf{T}} \tag{3.23}$$

can be used to determine a unique error pattern of Mannheim weight one. However, for the AWGN channel, codewords with a Mannheim distance two to the received word can have a smaller squared Euclidean distance than the unique codeword with a Mannheim distance one.

The presented decoding method comprises two steps. Within the first step, the syndrome is calculated. The syndrome determines a list of error patterns with weight one and two. Each error pattern $\boldsymbol{e}$ leads to a candidate codeword $\tilde{\boldsymbol{y}} - \boldsymbol{e}$. Therefore, we are able to construct a list $\mathcal{L}$ of candidate codewords, which includes all codewords with a Mannheim distance two and the unique codeword with distance one. Within the second decoding step, we select the codeword $\boldsymbol{c} \in \mathcal{L}$ that minimizes the squared Euclidean distance to the received vector $\boldsymbol{y}$, i.e.

$$\hat{\boldsymbol{c}} = \underset{\boldsymbol{c}\in\mathcal{L}}{\operatorname{argmin}}\, ||\boldsymbol{y} - \boldsymbol{c}||^2. \tag{3.24}$$

We illustrate this concept in Example 3.2 for codes over Gaussian integers.

Example 3.2 *We use the field* $\mathcal{G}_{13}$ *from Example 3.1 and the primitive element* $\alpha = 1 + \mathrm{i}$. *A code of length* $N_\mathrm{c} = 3$ *over this field has the parity-check matrix* $\boldsymbol{P} = [1, 1+\mathrm{i}, 2\mathrm{i}]$. *This code has a minimum Mannheim distance* $\mathrm{dist}_\mathrm{M} = 3$ *and is able to correct any single error from the set* $\{1, -1, \mathrm{i}, -\mathrm{i}\}$. *For example, the received vector* $\tilde{\boldsymbol{y}} = [2, 1+\mathrm{i}, -1]$ *leads to the syndrome* $s = \boldsymbol{P}\tilde{\boldsymbol{y}}^{\mathsf{T}} = 2$. *Using the one-error decoding method presented in [Hub94b],* *we obtain an error vector* $\boldsymbol{e}_1 = [0, 0, -\mathrm{i}]$ *and the estimated codeword* $\boldsymbol{c}_1 = [2, 1+\mathrm{i}, -1+\mathrm{i}]$ *which has a Mannheim distance one to the received vector. However, we can obtain other codeword candidates. Table 3.1 contains all error patterns of weight two for the syndrome value 2. For instance, with the error pattern* $\boldsymbol{e}_2 = [0, \mathrm{i}, \mathrm{i}]$, *we also obtain the syndrome* $\boldsymbol{P}\boldsymbol{e}_2^{\mathsf{T}} = 2$. *The corresponding codeword* $\boldsymbol{c}_2 = \tilde{\boldsymbol{y}} - \boldsymbol{e}_2 = [2, 1, -1-\mathrm{i}]$ *has a Mannheim distance* $\mathrm{dist}_{\mathrm{M}}(\tilde{\boldsymbol{y}}, \boldsymbol{c}_2) = 2$. *Overall, we can obtain a list with seven codewords using all error patterns from Table 3.1 for the same syndrome.*

Note that for codes over Eisenstein integers, there are six errors of weight one, i.e. $e_j \in \{\pm 1, \pm\omega, \pm(1+\omega)\}$. We illustrate the concept applied to Eisenstein integers in Example 3.3.

Example 3.3 *We use the field* $\mathcal{E}_{19}$ *and the primitive element* $\alpha = 2$. *A code of length* $N_c = 3$ *over this field has the parity-check matrix* $\boldsymbol{P} = [1, 2, -1-2\omega]$. *This code is able to correct any single error from the set* $\{\pm 1, \pm\omega, \pm(1+\omega)\}$, *i.e. one error in one position to a next neighbor. For example, the received vector* $\tilde{\boldsymbol{y}} = [1, -1, 1+2\omega]$ *leads to the syndrome* $s = \boldsymbol{P}\tilde{\boldsymbol{y}}^{\mathsf{T}} = 2$. *Using the one-error decoding method presented in [Hub94a], we obtain an error vector* $\boldsymbol{e}_1 = [0, 1, 0]$ *and the estimated codeword* $\boldsymbol{c}_1 = [1, -2, 2\omega]$. *However, we can obtain other codeword candidates. Table 3.2 contains all error patterns of weight two for the syndrome value 2. For instance, with the error pattern* $\boldsymbol{e}_2 = [0, -1, 1]$, *we also obtain the syndrome* $\boldsymbol{P}\boldsymbol{e}_2^{\mathsf{T}} = 2$. *The corresponding codeword is* $\boldsymbol{c}_2 = \tilde{\boldsymbol{y}} - \boldsymbol{e}_2 = [1, 0, 2\omega]$. *Overall, we can obtain a list with ten codewords for the same syndrome.*

The lookup tables as presented in Tables 3.1 and 3.2 can be calculated by generating all error patterns of weight two, where the error patterns are ordered with respect to the syndrome $s = \boldsymbol{P}\boldsymbol{e}^{\mathsf{T}}$. Note that it is sufficient to store the syndrome values of one quadrant, because $-s = \boldsymbol{P}(-\boldsymbol{e}^{\mathsf{T}})$, $\mathrm{i}s = \boldsymbol{P}\left(\mathrm{i}\boldsymbol{e}^{\mathsf{T}}\right)$, $-\mathrm{i}s = \boldsymbol{P}\left(-\mathrm{i}\boldsymbol{e}^{\mathsf{T}}\right)$.

Next, we present results of Monte-Carlo simulations for transmission over the AWGN channel. The corresponding simulation results in terms of SER versus SNR are depicted in Figure 3.2. We consider the constellations $\mathcal{E}_{19}$ and $\mathcal{G}_{17}$ with coded and uncoded transmission. For the coded transmission, we use OMEC codes of length $N_c = 3$ with ML decoding and the proposed list-decoding method of the previous section.

Table 3.2 Syndrome lookup table for the Eisenstein code from Example 3.3

syndrome s	1	$1+\omega$	2
error vector $\boldsymbol{e}$	$[0, 0, -2\omega]$	$[0, 0, 2]$	$[0, 0, -2-2\omega]$
	$[0, -\omega, -1]$	$[0, \omega, 1+\omega]$	$[0, 1, 0]$
	$[0, 1+\omega, 1]$	$[0, 1, -1-\omega]$	$[0, -1, 1]$
	$[0, \omega, -1-\omega]$	$[0, -1, -\omega]$	$[-1-\omega, 0, -\omega]$
	$[0, -2-\omega, 0]$	$[0, -1-2\omega, 0]$	$[-\omega, 0, \omega]$
	$[-1-\omega, 0, \omega]$	$[-1-\omega, 1+\omega, 0]$	$[1+\omega, 1+\omega, 0]$
	$[\omega, 0, 1+\omega]$	$[-\omega, 0, -1]$	$[\omega, \omega, 0]$
	$[1, 0, 0]$	$[1+\omega, 0, 0]$	$[-1, -1-\omega, 0]$
	$[-1, 1, 0]$	$[-1, 0, \omega]$	$[2, 0, 0]$

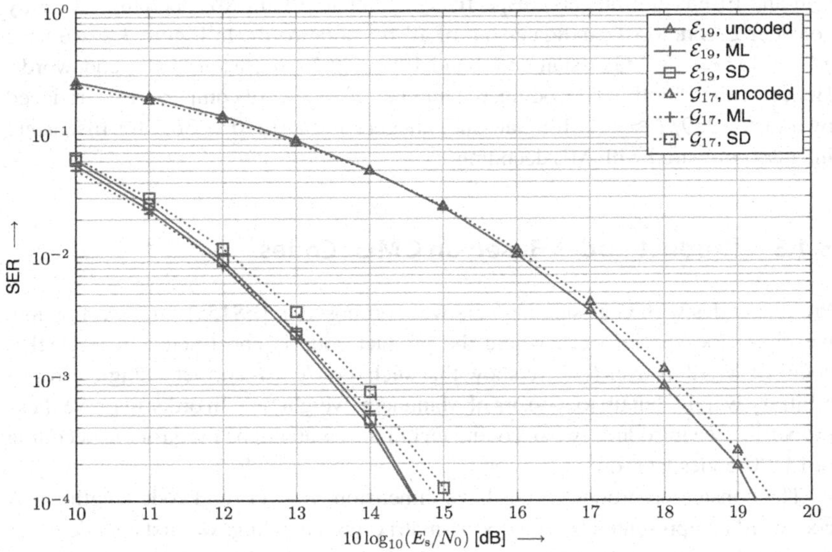

Figure 3.2 Simulation results in terms of SER versus SNR for transmission over the AWGN channel, results for Eisenstein integers ($\mathcal{E}_{19}$) and Gaussian integers ($\mathcal{G}_{17}$) with coded and uncoded transmission

Note that for the uncoded transmission, the Eisenstein constellation slightly outperforms the Gaussian constellation, despite the fact that the Eisenstein constellation contains more signal points. This performance gain results from the denser packing and the lower variance of the Eisenstein integers. Similarly, for coded transmission the OMEC code over Eisenstein integers outperforms the code over Gaussian integers with ML decoding.

The proposed suboptimal decoding (SD) results in near-ML performance with Eisenstein integers, whereas there is a small performance loss with Gaussian integers. The improved performance with Eisenstein integers comes at the cost of a higher decoding complexity with the suboptimal decoding due to the larger list size. For Gaussian integers, there are four errors of weight one, i.e. $e_j \in \{1, -1, \mathrm{i}, -\mathrm{i}\}$, whereas for Eisenstein integers we have $e_j \in \{\pm 1, \pm\omega, \pm(1+\omega)\}$. Hence, for Eisenstein integers we can generate more codeword candidates, which improves the decoding performance but results in a higher computational complexity. For the considered codes, we have list sizes $L = 7$ and $L = 10$ for the codes over Gaussian and Eisenstein integers, respectively. Compared with ML decoding, we only consider L codeword candidates out of all p^K codewords. In Figure 3.2, we have $p^K = 17^2 = 289$ (Gaussian code) and $19^2 = 361$ (Eisenstein code) codewords. Hence, with suboptimal decoding, the number of codeword comparisons is reduced by a factor $289/7 \approx 41.3$ for Gaussian integers and $361/10 = 36.1$ for Eisenstein integers compared with ML decoding.

3.1.3 Product Codes Based on OMEC Codes

Product codes with Gaussian integers were proposed in [FS15a]. In the following, we revise the special case, where the product code is constructed from OMEC codes of length n over $\mathcal{G}_p$. We show that such a code can correct a single error of arbitrary weight, but up to n errors of Mannheim weight one. In order to prove these properties, we introduce two decoding techniques that will be used for the decoding of LDPC codes later on.

The considered product codes have dimension $(n-1)^2$ and code length n^2. A codeword is represented by an $(n \times n)$-matrix. For encoding, we first encode $n-1$ codewords of the OMEC code and store these codewords column-wise into the first $n-1$ columns of the codeword matrix. Then, we use the OMEC code n-times to encode each row of the matrix. The resulting code has the following properties:

Proposition 3.1 *A product code constructed from OMEC codes of length n has a minimum Hamming distance of at least four and a minimum Mannheim distance of at least six. It can correct*

a) a single error of arbitrary Mannheim weight
b) a pattern of up to n errors with Mannheim weight one that occur either in different rows or different columns
c) two errors with Mannheim weight one in arbitrary error positions

Proof. According to [FS15a], the minimum Mannheim distance of a linear code over $\mathcal{G}_p$ is equivalent to the minimum Mannheim weight of a non-zero codeword. Similarly, the minimum Hamming distance of a linear code over $\mathcal{G}_p$ is equivalent to the minimum Hamming weight of a non-zero codeword. A non-zero codeword of the product code has at least one non-zero column. This column is a codeword of the OMEC code and has minimum Hamming distance $\text{dist}_\text{H} = 2$ and at least two non-zero elements. However, each non-zero element of this column results in a non-zero row. Each non-zero row is again a codeword of the OMEC code and has at least Hamming weight $\text{dist}_\text{H} = 2$ and Mannheim weight $\text{dist}_\text{M} = 3$. Hence, the minimum Hamming distance of the product code is at least four and the minimum Mannheim distance six.
Case a) Consider a single error of arbitrary Mannheim weight in the ith column and the jth row of the codeword matrix. Note that each column and row codeword has Hamming distance two and can detect any single error of arbitrary Mannheim weight, i.e. any single error results in an non-zero syndrome. Calculating the syndrome values for each row and each column, we can determine the error position, i.e. the ith column syndrome and the jth row syndrome are non-zero syndromes. The code symbol in the error position is considered as an erasure. Its value can be determined as shown in (3.18) for the error position using either the ith column or the jth row.
Case b) Consider an error pattern of errors with Mannheim weight one that occur either in different rows or different columns. For decoding we first calculate the syndrome for each row and each column. Next we determine the number i of non-zero syndromes for the rows and the number j of non-zero syndromes for the columns. Note that each row or column that is affected by a single error results in a non-zero syndrome, but multiple errors in the same OMEC codeword may result in a zero syndrome. Assume that $i \leq n$ errors occur in different rows. This results in i non-zero syndrome values for the rows. We have $j \leq i$ for the number of non-zero syndrome values for the columns, because multiple errors in columns may result in non-zero syndrome values. Likewise, if $j \leq n$ errors occur in different columns, we have $i \leq j$. Hence, we can correct all errors in different rows or different columns by comparing i with j and applying syndrome decoding along the dimension with the larger value. For $i = j$, we can decode either row- or column-wise.

Case c) Note that any two errors are either in independent rows or independent columns. Hence, c) follows from b). □

3.1.4 LDPC Codes based on OMEC Codes

In the following, we propose a construction of LDPC codes over Gaussian integers. Similar, non-binary LDPC codes were considered in [SF05, dPBZB12, THBN15]. In [SF05], LDPC codes over rings were proposed that use a mapping onto PSK. LPDC codes over lattices were investigated in [dPBZB12, THBN15]. We propose a construction for LDPC codes, that uses the complex-valued modulo function (3.4) as a mapping onto finite Gaussian integer fields. The sparse parity-check matrix is constructed from OMEC codes. Moreover, we introduce a new channel model for codes over Gaussian integers, the one Mannheim error channel, and study its channel capacity. This channel can be considered as a first order approximation of the AGWN channel with hard-decision detection (cf. (3.21)) of the transmitted symbol. For the one Mannheim error channel, we assume that only errors to nearest neighbors in the signal constellation occur. We demonstrate that the proposed codes can efficiently correct many errors of Mannheim weight one.

Channel model

Consider transmissions of complex valued q-ary symbols over an AWGN channel. For such a transmission, the symbol error probability is dominated by symbol errors to the nearest neighbors within the signal constellation, i.e. to the symbols that have the smallest Euclidean distance to the actually transmitted symbols. For a Gaussian integer constellation, these errors have the Mannheim distance $\text{dist}_{\text{M}} = 1$.

In the following, we consider a simplified channel model, which can be considered as a first order approximation of the AGWN channel model with hard decision detection of the transmitted symbol at the receiver, where we assume that only errors to nearest neighbors occur. The proposed channel model is a discrete and memoryless channel, which considers only errors of Mannheim weight one.

The one Mannheim error channel is a discrete and memoryless channel defined as

$$\boldsymbol{y} = \lfloor \boldsymbol{c} + \boldsymbol{e} \rceil \tag{3.25}$$

where $\boldsymbol{c}$ denotes the transmitted codeword with $c_i \in \mathcal{G}_p$. $\boldsymbol{y}$ is the received vector. $\boldsymbol{e}$ corresponds to the error vector with elements $e_i \in \{0, \pm 1, \pm \text{i}\}$. Addition is performed element-wise. Errors occur independently with symbol error probability ϵ. All error symbols $\{\pm 1, \pm \text{i}\}$ are equally likely, i.e.

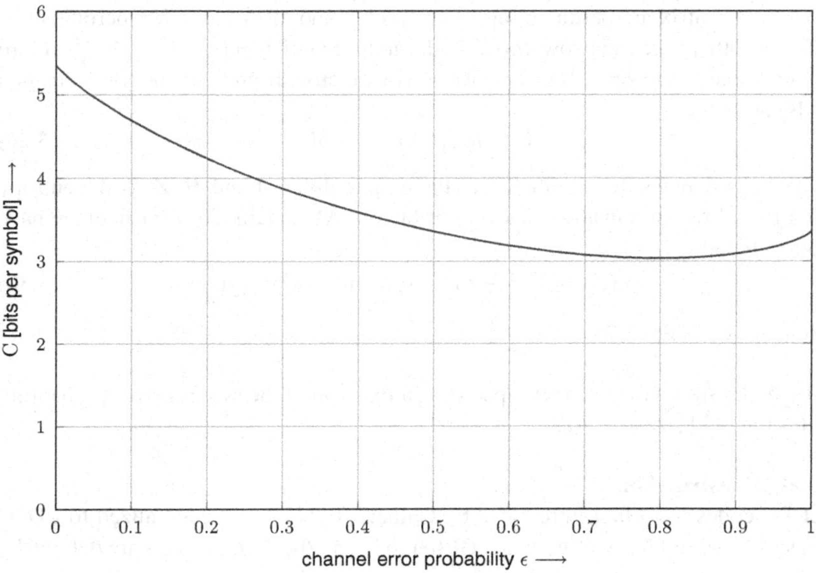

Figure 3.3 Channel capacity for the one Mannheim error channel for $p = 41$

$$P(e_i = 0) = 1 - \epsilon \tag{3.26}$$

$$P(e_i = 1) = P(e_i = -1) = P(e_i = \mathrm{i}) = P(e_i = -\mathrm{i}) = \frac{\epsilon}{4}. \tag{3.27}$$

Proposition 3.2 *The channel capacity* C *of the one Mannheim error channel in bits per transmitted symbol is*

$$C = \log_2(p) + (1 - \epsilon) \cdot \log_2(1 - \epsilon) + \epsilon \cdot \log_2(\frac{\epsilon}{4}), \tag{3.28}$$

where p denotes the size of the finite Gaussian integer field $\mathcal{G}_p$.

Proof. By definition, the one Mannheim error channel is a discrete memoryless channel. A discrete memoryless channel can be characterized by a transition matrix that contains all transmission probabilities from an input symbol $\mathcal{C}$ to the channel output $\mathcal{Y}$. The one Mannheim error channel is symmetric, because all rows of the transition matrix are permutations of each other and all columns are permutations of each other, i.e. each row contains the non-zero elements $\epsilon, \frac{\epsilon}{4}, \frac{\epsilon}{4}, \frac{\epsilon}{4}, \frac{\epsilon}{4}$ and all other elements are zero. The capacity of a symmetric discrete memoryless channel is [Gal68]

$$C = \log_2(|\mathcal{Y}|) - H(\mathbf{Z}), \tag{3.29}$$

where $|\mathcal{Y}|$ denotes the cardinality of the output alphabet and $H(\mathbf{Z})$ is the entropy of a row $\mathbf{Z}$ of the transition matrix. For the one Mannheim error channel we have $|\mathcal{Y}| = p$ and

$$H(\mathbf{Z}) = -(1-\epsilon)\log_2(1-\epsilon) - \epsilon \cdot \log_2(\frac{\epsilon}{4}). \tag{3.30}$$

Hence, we obtain (3.28). □

Figure 3.3 shows the channel capacity as a function of the symbol error probability ϵ for the field size $p = 41$.

Code Construction

LDPC codes were first introduced by Gallager [Gal63] and generalized to LDPC codes over non-binary finite fields GF(q) in [DM98b]. LDPC codes are defined by a sparse parity-check matrix $\boldsymbol{P}$ which contains mostly zeros and a small number of non-zero elements. Regular LDPC codes are defined by the triple $[N, J, K]$ with $J \geq 2$ and $K > J$, where N is the code length, J is the number of non-zero elements per column, and K is the number of non-zero elements per row of $\boldsymbol{P}$. Such an LDPC code rate is $R = 1 - \frac{J}{K}$.

Similarly, the proposed LDPC codes over $\mathcal{G}_p$ can be defined by the quadruple $[N_c, J, K, p]$, where p denotes the size of the field. The parity-check matrix $\boldsymbol{P}$ can be constructed from J sub-matrices. The first $\frac{N_c}{K} \times N_c$ sub-matrix $\boldsymbol{P}_1$ is defined as

$$\boldsymbol{P}_1 = \begin{pmatrix} \alpha^0 \dots \alpha^{K-1} & 0 \dots & 0 & 0 & \\ 0 \dots & 0 & \alpha^0 \dots \alpha^{K-1} & 0 & \dots \\ 0 \dots & 0 & 0 \dots & 0 & \alpha^0 \\ & & \vdots & & \ddots \end{pmatrix}, \tag{3.31}$$

where α is a primitive element of $\mathcal{G}_p$. The sub-matrices $\boldsymbol{P}_j$ for $j = 2, \dots, J$ are generated by random column permutations of $\boldsymbol{P}_1$. Finally, we get $\boldsymbol{P}$ by

$$P = \begin{pmatrix} P_1 \\ P_2 \\ \vdots \\ P_J \end{pmatrix}. \tag{3.32}$$

Furthermore, with codes over Gaussian integers we have the restriction $K \leq (p-1)/4$, because we apply syndrome decoding for the check equations. The code rate is $R = (1 - \frac{J}{K})$ or $R \cdot \log_2(p)$ in bpcu.

Decoding

We propose a simple non-probabilistic iterative decoding algorithm similar to Gallager's decoding algorithm A [Gal63, UR08], which is a bit-flipping procedure for binary LDPC codes. We consider transmission over the one Mannheim error channel.

Each row of the parity-check matrix P is a check equation for K code symbols. That is, the K code symbols corresponding to the non-zero elements of a row of P form a codeword of an OMEC code. We consider two methods to decode such a check equation, i.e. we use erasure and syndrome decoding steps similar to the decoding of the product code discussed in Section 3.1.3. Both decoding rules are executed alternately and iteratively for a certain number of iterations.

Erasure decoding: Consider the received symbol y_i. This symbol is connected to J check equations by the J non-zero entries in the ith column of P. Using the J check equations, we can determine the value of the code symbol v_i based on the other $K-1$ received symbol values that correspond to non-zero entries in the same row of P. If more than $\lceil \frac{J}{2} \rceil$ check equations result in the same value $\hat{c}_i$ for the ith symbol we change the value of y_i to $\hat{c}_i$, otherwise we proceed with an unaltered value for y_i. We may also restrict the value of $\hat{c}_i$ to the neighborhood of the received value y_i in the constellation $\mathcal{G}_p$. That is, we change the value of y_i only if $\text{dist}_M(y_i, \hat{c}_i) \leq e_{\max}$, where $e_{\max}$ is the maximum error weight that will be corrected. In particular, in the first decoding iteration we assume $e_{\max} = 1$, but this value is increased with the number of iterations.

Syndrome decoding: With erasure decoding, we determine the value of an erased or unreliable code symbol, whereas with syndrome decoding we determine an error value $e_i \in \{\pm 1, \pm \mathrm{i}\}$. For each check equation with a non-zero syndrome value, we can determine the error location and an error value based on OMEC syndrome decoding.

We use a message passing procedure to determine which symbols should be altered. The error value e_i will be sent to the i_{th} message node. Moreover, all K symbols incident to a violated check equation receive a message that indicates this violation. Consider the ith symbol. If just one equation sends a non-zero error value, but J violation message are received, then we alter the symbol to

$$c_i = y_i - e_i \tag{3.33}$$

Similarly, if more than $\lceil \frac{J}{2} \rceil$ non-zero error values e_i are identical, we alter the symbol.

Simulation Results

We present simulation results for a $[10000, 3, 10, 41]$ code. This code has rate $R = 0.7$ or 3.75 bits per symbol.

Figure 3.4 shows the performance for the two proposed decode algorithms, where only erasure or syndrome decoding is used. For $\epsilon < 0.2$, syndrome decoding has a better performance than erasure decoding. Note that syndrome decoding is a unique feature of the used OMEC codes. The non-probabilistic decoding of an LDPC code with arbitrary q-ary entries in the parity-check matrix could only be based on the erasure decoding procedure.

The decoding performance can be improved, when both decoding procedures are used alternately. Figure 3.5 shows a significant performance gain by combining both algorithms. We applied the maximum error weight that will be corrected with $\text{dist}_{\text{M}}(y_i, \hat{c}_i) \leq e_{\max}$. For a) the limit is $e_{\max} = 2$. Most of the remaining errors for $\epsilon < 0.1$ are of weight higher than $e_{\max} = 2$. b) follows an approach, where we increment e_{max} in every third iteration, starting with $e_{\max} = 1$. This decoding approach achieves the best performance for a low channel error probability. c) shows the performance for a product code as described in Section 3.1.3. The code is based on the OMEC code with elements from $p = 41$, has length $N_c = 36$ and rate $R = 0.7$, which is the same as that for a) and b). For $\epsilon > 0.13$, the product code shows the best performance.

3.2 Four-Dimensional Signal Constellations

In this section, we first provide a brief introduction to Lipschitz and Hurwitz integers and their properties. Subsequently, a new construction method for Lipschitz- and Hurwitz-integer signal constellations is given. We compare the proposed constellations to the Lipschitz-integer constellations from [FS15a] and the Gaussian-integer

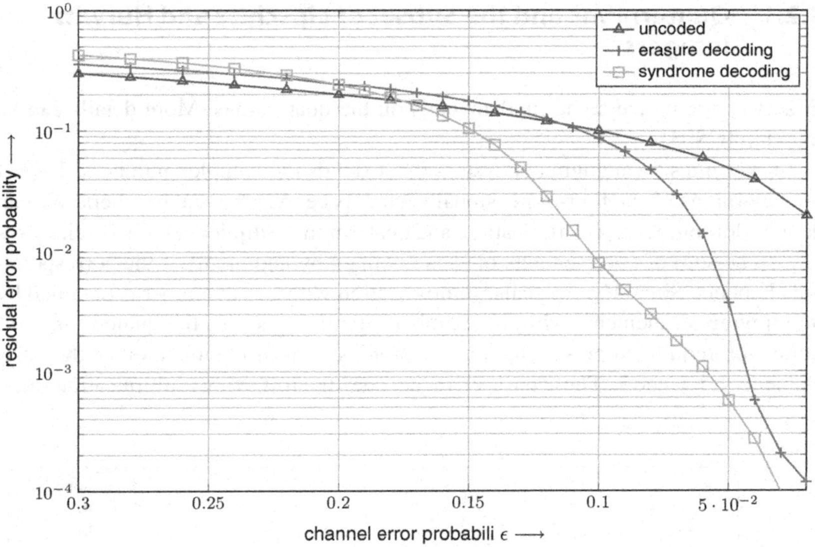

Figure 3.4 Comparison of erasure and syndrome decoding

constellations from [Hub94b, FGS13] based on the constellation figure of merit [FW89], which is a measure for the error performance for transmission over the AWGN channel, as well as by Monte-Carlo simulations.

Furthermore, two set-partitioning techniques for Lipschitz and Hurwitz constellations are presented. The method described in Section 3.2.4 aims to reduce the search space for the symbol detection, where dependencies between dimensions are exploited. Since the search space is reduced, the computational complexity is also reduced. Simulation results for the AWGN channel demonstrate that the new Hurwitz constellations with suboptimal detection can outperform conventional two-dimensional constellations with ML detection for high-SNR values. The second partitioning method aims to find subsets with a large minimum squared Euclidean distance. It was originally presented in [FSS14] for Gaussian integers. In Section 3.2.5, this approach is generalized to the proposed Hurwitz integer constellations. The subsets are additive subgroups and they can have much higher minimum Euclidean distances than the original constellations. This technique enables multilevel coding over Hurwitz constellations, which is discussed in Section 3.2.6.

3.2.1 Quaternions and the Subset of Lipschitz and Hurwitz Integers

First, we summarize some fundamentals on the quaternions. More details can be found in [CS03].

Quaternions form a number system that extends the complex numbers. The set of quaternions $\mathbb{H}$ is a four-dimensional vector space over the real numbers. We can define addition, scalar multiplication, and quaternion multiplication for the elements of $\mathbb{H}$. The sum of two elements of $\mathbb{H}$ is obtained by adding all of the corresponding elements. Similarly, the scalar multiplication with a real number is obtained by multiplying all elements with this number. The definition of the quaternion multiplication requires representing a quaternion as a linear combination of the basis elements 1, i, j, and k. These basis elements obey the following multiplication rules

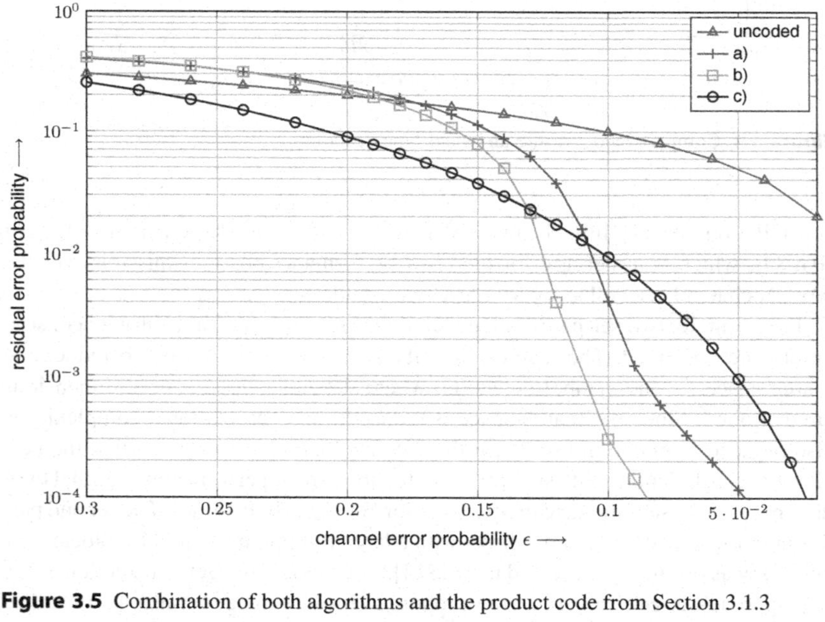

Figure 3.5 Combination of both algorithms and the product code from Section 3.1.3

$$
\begin{aligned}
\mathrm{i}^2 &= \mathrm{j}^2 = \mathrm{k}^2 = -1, \\
\mathrm{ij} &= \mathrm{k} = -\mathrm{ji}, \\
\mathrm{jk} &= \mathrm{i} = -\mathrm{kj}, \\
\mathrm{ki} &= \mathrm{j} = -\mathrm{ik}\,.
\end{aligned} \tag{3.34}
$$

A quaternion $a \in \mathbb{H}$ is defined as

$$a = a_1 + a_2\mathrm{i} + a_3\mathrm{j} + a_4\mathrm{k}\,, \tag{3.35}$$

where $a_1, a_2, a_3, a_4 \in \mathbb{R}$. Quaternion multiplication of two quaternions $a = a_1 + a_2\mathrm{i} + a_3\mathrm{j} + a_4\mathrm{k}$ and $b = b_1 + b_2\mathrm{i} + b_3\mathrm{j} + b_4\mathrm{k}$ is defined as

$$
\begin{aligned}
a \cdot b = &(a_1b_1 - a_2b_2 - a_3b_3 - a_4b_4) + \\
&(a_1b_2 + a_2b_1 + a_3b_4 - a_4b_3)\mathrm{i} + \\
&(a_1b_3 - a_2b_4 + a_3b_1 + a_4b_2)\mathrm{j} + \\
&(a_1b_4 + a_2b_3 - a_3b_2 + a_4b_1)\mathrm{k}\,.
\end{aligned} \tag{3.36}
$$

The conjugate of a quaternion a is

$$a^* = a_1 - a_2\mathrm{i} - a_3\mathrm{j} - a_4\mathrm{k} \tag{3.37}$$

and its norm is

$$N(a) = aa^* = a_1^2 + a_2^2 + a_3^2 + a_4^2. \tag{3.38}$$

Multiplication of two quaternions is not commutative. However, we have $N(a) = aa^* = a^*a$. The inverse for $a \neq 0$ is $\frac{a^*}{||a||^2}$, i.e. $\frac{aa^*}{||a||^2} = 1$. A quaternion is called primitive if the largest common divisor of its components is one. Furthermore, the squared Euclidean distance of two quaternions or complex-valued numbers is

$$\mathrm{dist}_\mathrm{E}(a, b) = N(a - b) \tag{3.39}$$

and the minimum squared Euclidean distance of the constellation $\mathcal{X}$ is

$$\delta^2 = \min_{\substack{a,b\in\mathcal{X} \\ a\neq b}} \mathrm{dist}_\mathrm{E}(a, b). \tag{3.40}$$

The set of Hurwitz integers $\mathscr{H}$ is a subset of the quaternions $\mathbb{H}$, where a Hurwitz integer is a quaternion whose components are all integers or all half-integers (halves of an odd integer), i.e.

$$\mathscr{H} = \left\{ a_1 + a_2\mathrm{i} + a_3\mathrm{j} + a_4\mathrm{k} \,\middle|\, a_1, a_2, a_3, a_4 \in \mathbb{Z} \text{ or } a_1, a_2, a_3, a_4 \in \mathbb{Z} + \frac{1}{2} \right\} \subset \mathbb{H}. \tag{3.41}$$

The set of Lipschitz integers $\mathscr{L}$ is a subset of $\mathscr{H}$, where a Lipschitz integer is a quaternion whose components are all integers, i.e.

$$\mathscr{L} = \left\{ a_1 + a_2\mathrm{i} + a_3\mathrm{j} + a_4\mathrm{k} \middle| a_1, a_2, a_3, a_4 \in \mathbb{Z} \right\} \subset \mathscr{H}. \tag{3.42}$$

As with Gaussian integers, we use $\lfloor \cdot \rceil$ to denote rounding, i.e. for a real number a_j, $\lfloor a_j \rceil$ yields the integer closest to it, where we round half-way values towards plus infinity. For a quaternion $a = a_1 + a_2\mathrm{i} + a_3\mathrm{j} + a_4\mathrm{k}$, we have $\lfloor a \rceil = \lfloor a_1 \rceil + \lfloor a_2 \rceil \mathrm{i} + \lfloor a_3 \rceil \mathrm{j} + \lfloor a_4 \rceil \mathrm{k}$. Let λ be a Lipschitz integer with non-zero norm. Similar to Gaussian integers [Hub94b], we can define the modulo function of a Lipschitz integer a as

$$\mu_\lambda(a) = a \bmod \lambda = a - \left\lfloor \frac{a\lambda^*}{N(\lambda)} \right\rceil \lambda = a - \left\lfloor a\lambda^{-1} \right\rceil \lambda, \tag{3.43}$$

where $\mu_\lambda(a)$ is the remainder of the Euclidean division of a by λ. The modulo operation (3.43) performs a reduction from a particular point in four-dimensional space to its congruent one located in the Voronoi cell of the Lipschitz integers (w.r.t. the origin), see also [SF15, Ste19]. In particular, that Voronoi cell (a four-dimensional hypercube) is scaled and rotated in four-dimensional space as defined by the Lipschitz integer λ.

The modulo function (3.43) can be implemented as a Euclidean division algorithm similar to the modulo operation for Eisenstein integers proposed in [SYHS13]. However, Eisenstein integers are a Euclidean domain which ensures that the norm of the remainder is smaller than the norm of the divisor. However, the set of Lipschitz integers is not a Euclidean domain. Consequently, the result of a Euclidean division algorithm may not be unique. In (3.43) uniqueness is achieved by the rounding, where we round half-way values towards plus infinity. We demonstrate this in the following example and provide a general proof for uniqueness subsequently.

Example 3.4 *Consider the Lipschitz integers* $a = -1 - 2\mathrm{i} - 2\mathrm{j} - \mathrm{k}$ *and* $b = 2 - \mathrm{i} - 2\mathrm{j} - \mathrm{k}$. *These numbers are congruent with respect to the division by* $\lambda = 3 + \mathrm{i}$, *i.e.* $b = a + \lambda$. *Moreover, all three Lipschitz integers have the norm* $N(\lambda) = N(a) = N(b) = 10$. *The modulo function provides the unique result* $\mu_\lambda(a) = \mu_\lambda(b) = a$. *Consider for instance the quotient* $a\lambda^{-1} = -\frac{1}{2} - \frac{\mathrm{i}}{2} - \frac{\mathrm{j}}{2} - \frac{\mathrm{k}}{2}$. *The value* $-\frac{1}{2}$ *will be rounded to zero, whereas* $\frac{1}{2}$ *will be rounded to one. Hence, we obtain the rounded quotient* $\left[a\lambda^{-1}\right] = 0$ *and* $\mu_\lambda(a) = a$. *On the other hand, for b we have the quotient* $b\lambda^{-1} = \frac{1}{2} - \frac{\mathrm{i}}{2} - \frac{\mathrm{j}}{2} - \frac{\mathrm{k}}{2}$ *which will be rounded to* $\left[b\lambda^{-1}\right] = 1$. *The modulo operation results in* $\mu_\lambda(b) = b - \lambda = a$.

Now, consider the class $\mathcal{K}_\lambda(z)$ of a Lipschitz integer z which is the set of all congruent numbers z' such that $z' = z + c\lambda$ for some Lipschitz integer c.

Proposition 3.3 *The class* $\mathcal{K}(z)$ *has a unique representative* z *with* $z = \mu_\lambda(z)$.

Proof. We proof the proposition by contradiction. We assume that beside z there exists a second element $z' \in \mathcal{K}(z)$ that satisfies $z' = \mu_\lambda(z') \neq z$. Note that $z = \mu_\lambda(z)$ implies a rounded quotient $\left[z\lambda^{-1}\right] = 0$. Let a_1, a_2, a_3, a_4 be the components of the quaternion $z\lambda^{-1} = a_1 + a_2\mathrm{i} + a_3\mathrm{j} + a_4\mathrm{k}$. From $\left[z\lambda^{-1}\right] = 0$ follows that $|a_l| \leq \frac{1}{2}$ with equality only for $a_l = -\frac{1}{2}$. Similarly, $z' = \mu_\lambda(z')$ requires $\left[z'\lambda^{-1}\right] = 0$.

The congruence implies that $z' = z + c\lambda$ for some non-zero Lipschitz integer c. For z', we obtain the quotient $z'\lambda^{-1} = z\lambda^{-1} + c = a_1 + a_2\mathrm{i} + a_3\mathrm{j} + a_4\mathrm{k} + c$. The Lipschitz integer $c = c_1 + c_2\mathrm{i} + c_3\mathrm{j} + c_4\mathrm{k}$ has at least one non-zero integer component c_l. The condition $|a_l| \leq \frac{1}{2}$ implies $|c_l + a_l| \geq \frac{1}{2}$ with equality only for $c_l + a_l = \frac{1}{2}$. Consequently, the rounded quotient $z'\lambda^{-1}$ is non-zero which is a contradiction to $z' = \mu_\lambda(z')$. □

Moreover, note that the order of the operands in (3.43) matters since the multiplication of two quaternions is not commutative.

Example 3.5 *We consider the modulo function for the divisor* $\lambda = 3 + \mathrm{i}$. *The two products* $a = \mathrm{j}\lambda = 3\mathrm{j} - \mathrm{k}$ *and* $b = \lambda\mathrm{j} = 3\mathrm{j} + \mathrm{k}$ *result from* (3.34) *because* $\mathrm{ij} = k$ *and* $\mathrm{ji} = -k$. *These products show that the order matters. Moreover, we have* $\mu_\lambda(a) = 0$ *and* $\mu_\lambda(b) = -\mathrm{j} - \mathrm{k}$ *which illustrates that* $\lambda\mathrm{i}$ *is not divisible by* λ *when we multiply by* λ^{-1} *from the right.*

3.2.2 Construction of Finite Sets

Next, we present a new construction for Lipschitz- and Hurwitz-integer signal constellations. We denote the ring of integers modulo q by $\mathbb{Z}_q$, where for q prime, $\mathbb{Z}_q$ is a finite field. Furthermore, $\mathbb{Z}_q^2$ is the module $\mathbb{Z}_q^2 = \{[z_1, z_2] : z_1, z_2 \in \mathbb{Z}_q\}$, where scalar multiplication and vector addition are performed element-wise modulo q. The proposed four-dimensional signal constellations are obtained from the following mapping

$$\mathcal{L}_\lambda = \{\mu_\lambda(a + b\mathrm{j}) : a, b \in \mathbb{Z}_q, q = N(\lambda)\} \subset \mathscr{L}, \tag{3.44}$$

which provides the finite subset of Lipschitz integers. The corresponding coset of half-integers is obtained by

$$\mathcal{V}_\lambda = \{\mu_\lambda(h + w) : h \in \mathcal{L}_\lambda\} \subset \mathscr{H} \setminus \mathscr{L}, \tag{3.45}$$

with $w = \frac{1}{2} + \frac{1}{2}\mathrm{i} + \frac{1}{2}\mathrm{j} + \frac{1}{2}\mathrm{k}$. Finally, the union of both sets is the Hurwitz-integer signal constellation

$$\mathcal{H}_\lambda = \mathcal{L}_\lambda \cup \mathcal{V}_\lambda \subset \mathscr{H}. \tag{3.46}$$

Example 3.6 shows the construction method to obtain the constellation $\mathcal{H}_{2+\mathrm{i}}$.

Example 3.6 *We consider the proposed construction to obtain the Hurwitz constellation $\mathcal{H}_{2+\mathrm{i}}$ with $M_4 = 50$ points. First, to create the subset of Lipschitz integers according to* (3.44)*, we use $\lambda = 2 + \mathrm{i}$ with norm $q = N(2 + \mathrm{i}) = 5$, and thus $a, b \in \mathbb{Z}_5$. The modulo function* (3.43) *is applied to each of the $q^2 = 5^2 = 25$ elements of the set. For example, $a = 3$ and $b = 4$ are mapped to the Lipschitz integer $\mu_{2+\mathrm{i}}(3 + 4\mathrm{j}) = 1\mathrm{i} - 1\mathrm{j}$, given by*

$$\begin{aligned}
\mu_{2+\mathrm{i}}(3 + 4\mathrm{j}) &= 3 + 4\mathrm{j} \bmod 2 + \mathrm{i} \\
&= 3 + 4\mathrm{j} - \left\lfloor \frac{(3 + 4\mathrm{j})(2 - \mathrm{i})}{5} \right\rceil (2 + \mathrm{i}) \\
&= 3 + 4\mathrm{j} - \lfloor 1.2 - 0.6\mathrm{i} + 1.6\mathrm{j} + 0.8\mathrm{k} \rceil (2 + \mathrm{i}) \\
&= 3 + 4\mathrm{j} - (1 - 1\mathrm{i} + 2\mathrm{j} + 1\mathrm{k})(2 + \mathrm{i}) \\
&= 3 + 4\mathrm{j} - 3 + 1\mathrm{i} - 5\mathrm{j} + 0\mathrm{k} \\
&= 1\mathrm{i} - 1\mathrm{j}.
\end{aligned} \tag{3.47}$$

We utilize the subset of Lipschitz integers $\mathcal{L}_{2+\mathrm{i}}$ to create the corresponding coset of half-integers $\mathcal{V}_{2+\mathrm{i}}$ according to (3.45)*. Continuing the example for one point, we use $h = 1\mathrm{i} - 1\mathrm{j} \in \mathcal{L}_{2+\mathrm{i}}$ from* (3.47)*. Then,* (3.45) *is calculated as*

$$\begin{aligned}
\mu_{2+\mathrm{i}}(1\mathrm{i} - 1\mathrm{j} + w) &= 1\mathrm{i} - 1\mathrm{j} + w \bmod 2 + \mathrm{i} \\
&= 1\mathrm{i} - 1\mathrm{j} + w - \left\lfloor \frac{(1\mathrm{i} - 1\mathrm{j} + w)(2 - \mathrm{i})}{5} \right\rceil (2 + \mathrm{i}) \\
&= 1\mathrm{i} - 1\mathrm{j} + w - \lfloor 0.5 + 0.5\mathrm{i} - 0.3\mathrm{j} + 0.1\mathrm{k} \rceil (2 + \mathrm{i}) \\
&= 1\mathrm{i} - 1\mathrm{j} + w - (1 + 1\mathrm{i} + 0\mathrm{j} + 0\mathrm{k})(2 + \mathrm{i}) \\
&= 1\mathrm{i} - 1\mathrm{j} + 0.5 + 0.5\mathrm{i} + 0.5\mathrm{j} + 0.5\mathrm{k} - 1 - 3\mathrm{i} + 0\mathrm{j} + 0\mathrm{k} \\
&= -0.5 - 1.5\mathrm{i} - 0.5\mathrm{j} + 0.5\mathrm{k}. \qquad (3.48)
\end{aligned}$$

The union of both sets according to (3.46) *gives the set $\mathcal{H}_{2+\mathrm{i}}$ with $M_4 = 2q^2 = 25 + 25 = 50$ elements.*

We will show that the subset of Lipschitz integers $\mathscr{L}$ is isomorphic to the vector space or module $\mathbb{Z}_q^2$, which corresponds to a finite two-dimensional Euclidean geometry [LC04]. This definition of the finite set of Lipschitz integers differs from the construction presented in [FS15a], where the set of Lipschitz integers is isomorphic to the integer ring $\mathbb{Z}_q$. Moreover, we will demonstrate that the new construction provides denser packings, which results in higher constellation figure of merit (CFM) [FW89] values. Likewise, the construction of the finite sets of Hurwitz integers $\mathcal{H}_\lambda$ differs from the construction presented in [Gü13, GH14, Gü18]. The proposed construction results in signals sets with $2q^2$ elements, i.e. the number of elements is even. The construction presented in [Gü13, GH14, Gü18] results in signal sets, where the number of elements is odd.

We prove some properties of the sets $\mathcal{L}_\lambda$ and $\mathcal{H}_\lambda$ in the following proposition. In particular, we consider the special case where λ is a Lipschitz integer of the form $\lambda = a_1 + a_2\mathrm{i}$, i.e. λ is a Gaussian integer.

Proposition 3.4 *Let λ be a Lipschitz integer of the form $\lambda = a_1 + a_2\mathrm{i}$ with $q = N(\lambda)$, where q is a prime of the form $q \mod 4 = 1$, then the following holds*

- *$\mathcal{L}_\lambda$ is a vector space with q^2 elements isomorph to the finite Euclidean geometry $\mathbb{Z}_q^2$.*
- *$\mathcal{H}_\lambda$ is an additive group with $2q^2$ elements.*

Proof. In order to show that $\mathcal{L}_\lambda$ is isomorphic to the Euclidean geometry, we have to prove that the mapping $\mu_\lambda(\cdot)$ is bijective. Note that we can write any quaternion as $a_1 + a_2\mathrm{i} + a_3\mathrm{j} + a_4\mathrm{k} = \alpha + \beta\mathrm{j}$ with the complex numbers $\alpha = a_1 + \mathrm{i}a_2$ and $\beta = a_3 + a_4\mathrm{i}$, because $\mathrm{ij} = \mathrm{k}$. For Lipschitz integers of the form $\lambda = a_1 + a_2\mathrm{i}$, i.e. λ is a Gaussian integer, the mapping $\mu_\lambda(a + b\mathrm{j}) = \alpha + \beta\mathrm{j}$ can be calculated independently as $\alpha = \mu_\lambda(a)$ and $\beta = \mu_\lambda(b)$. Furthermore, with $q = N(\lambda)$, where q is a prime of the form $q \mod 4 = 1$, α and β are elements of the finite Gaussian integer field. Hence, the inverse mapping can also be calculated independently as $a = \mu_\lambda^{-1}(\alpha)$ and $b = \mu_\lambda^{-1}(\beta)$ with

$$\mu_\lambda^{-1}(\gamma) \equiv \gamma(v\lambda^*) + \gamma^*(u\lambda) \mod q \tag{3.49}$$

as shown in [Hub94b], where u, v are parameters that can be calculated with the Euclidean algorithm. The cardinality of $\mathcal{L}_\lambda$ is q^2, because $\mathbb{Z}_q^2$ has q^2 elements.

$\mathcal{L}_\lambda$ is an additive group because it is a vector space. $\mathcal{V}_\lambda$ is a coset of $\mathcal{L}_\lambda$ with q^2 elements. Consequently, $\mathcal{H}_\lambda = \mathcal{L}_\lambda \cup \mathcal{V}_\lambda$ is an additive group with $2q^2$ elements. □

Note that the proof also holds for cases where $q = N(\lambda)$ is not a prime, but the inverse mapping exists [FGS13]. In this case, $\mathbb{Z}_q$ is a ring and $\mathbb{Z}_q^2$ a module. Furthermore, the properties stated in Proposition 3.4 also hold for many cases, where λ is not a Gaussian integer. In particular, we verified numerically that all Lipschitz-integer sets listed in Table 3.3 are modules or vector spaces and the corresponding Hurwitz sets are additive groups.

3.2.3 CFM Comparison and ML Performance

In this subsection, we consider Lipschitz- and Hurwitz-integer signal constellations obtained by the construction method described in the previous subsection. The asymptotic symbol error probability of a signal constellation in the presence of AWGN depends on the minimum squared Euclidean distance of the constellation as well as the variance of its elements. The CFM combines these two measures and normalizes the variance to the variance of two-dimensional signals [FW89]. For a D-dimensional constellation $\mathcal{X}$, the CFM is defined as the ratio

$$\mathrm{CFM}(\mathcal{X}) = \frac{\mathrm{D}\delta^2}{2\sigma_x^2}. \tag{3.50}$$

Table 3.3 CFM for different Hurwitz-integer constellations (number of elements in the base ring, number of elements in the constellation, spectral efficiency, constellation figure of merit, moduli)

q	M_4	bpcu	CFM	λ
3	18	4.17	1.8947	$1+\mathrm{i}+\mathrm{j}$
5	50	5.64	1.1765	$2+\mathrm{i}$
6	72	6.17	0.9474	$2+\mathrm{i}+\mathrm{j}$
7	98	6.61	0.8485	$2+\mathrm{i}+\mathrm{j}+\mathrm{k}$
9	162	7.34	0.6626	$2+2\mathrm{i}+\mathrm{j}$
10	200	7.64	0.5882	$3+\mathrm{i}$
11	242	7.92	0.5432	$3+\mathrm{i}+\mathrm{j}$
13	338	8.40	0.4602	$3+2\mathrm{i}$
15	450	8.81	0.3991	$3+2\mathrm{i}+\mathrm{j}+\mathrm{k}$
17	578	9.18	0.3523	$4+\mathrm{i}$
18	648	9.34	0.3313	$3+2\mathrm{i}+2\mathrm{j}+\mathrm{k}$
19	722	9.50	0.3154	$4+\mathrm{i}+\mathrm{j}+\mathrm{k}$
21	882	9.79	0.2854	$4+2\mathrm{i}+\mathrm{j}$
22	968	9.92	0.2716	$4+2\mathrm{i}+\mathrm{j}+\mathrm{k}$
25	1250	10.29	0.2398	$4+3\mathrm{i}$
26	1352	10.40	0.2301	$4+3\mathrm{i}+\mathrm{j}$
27	1458	10.51	0.2221	$4+3\mathrm{i}+\mathrm{j}+\mathrm{k}$
29	1682	10.72	0.2068	$5+2\mathrm{i}$
31	1922	10.91	0.1934	$5+2\mathrm{i}+\mathrm{j}+\mathrm{k}$

A higher CFM leads to a lower SER for transmission over an AWGN channel. As mentioned at the beginning of this chapter, four-dimensional signal constellations can be applied in optical communications and with dual-polarized antennas in wireless communications, where the transmission is based on the in-phase and quadrature components of both polarizations [KA10, SFFF19, FSFF20]. This corresponds to the independent transmission of two complex-valued symbols at one time instance. In order to construct signal constellations from the Hurwitz integers that can be used with QAM, we use a_1 and a_3 as quadrature (real) components and a_2 and a_4 for the in-phase (imaginary) components. Hence, we have the straightforward mapping $x_1 = a_1 + a_2\mathrm{i}$ for the first transmitted symbol and $x_2 = a_3 + a_4\mathrm{i}$ for the second. This construction results in two projections $\mathcal{H}_\lambda \to \mathbb{C}$, where the first projection defines the signal set $\mathcal{T}_1$ for the first symbol and the second set $\mathcal{T}_2$ for the second symbol.

Table 3.3 presents some examples of Hurwitz constellations and their CFM values. In Examples 3.7 and 3.8, we compare the CFM values with those of the four-dimensional constellations $\mathcal{L}_\lambda^{(0)}$ presented in [FS15a] to demonstrate that these constellations are suitable for transmission over an AWGN channel. Moreover, the examples provide simulation results for transmission over the AWGN channel with 10^7 transmitted data symbols. At the receiver side, the ML detection rule is applied to retrieve the transmitted symbols. Signal constellations or sphere packings that have no algebraic group, ring or field property can reach higher CFM values, e.g. due to numerical optimized placement of points in space. Four-dimensional sphere packings for some cardinalities with up to $M_4 = 5698$ points are available in an online database at [Agr21]. For comparison, we use the set 14_50 with $M_4 = 50$ points and a CFM value of 1.3448. The Lipschitz constellations presented in [FS15a] have better CFM values than squared two-dimensional constellations (Gaussian

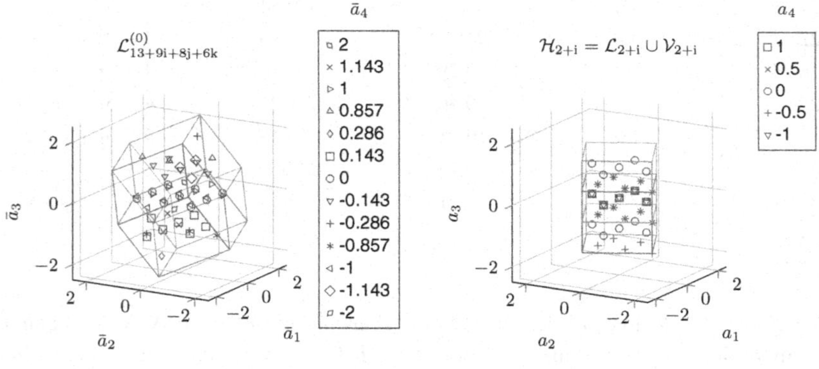

Figure 3.6 Illustration of four-dimensional constellations with $M_4 = 50$ points. Left: Lipschitz-integer constellation ($\mathcal{L}^{(0)}_{13+9i+8j+6k}$) from [FS15a]. Right: Hurwitz-integer constellation ($\mathcal{H}_{2+i}$). The constellations are drawn with a_1, a_2, and a_3 as Cartesian coordinates and a_4 is indicated by different marker shapes. For a visual comparison, the Lipschitz constellation is normalized to the same minimum squared Euclidean distance, which the Hurwitz constellation has by design, i.e. $\delta^2 = 1$. Integers and half integers are drawn in blue and red, respectively. The boundaries of the Voronoi regions (Voronoi cell of the Lipschitz integers (w.r.t. the origin) scaled by λ) are drawn as green or light grey colored and magenta or gray colored lines. Green or light gray colored lines connect vertices for $a_4 = 0.5$ and magenta (gray) colored lines for $a_4 = -0.5$ of the corresponding Lipschitz-integer Voronoi cell (points of the Voronoi cell, before (3.43) is performed)

integers [FGS13]) of the same size and obtain considerable asymptotic SNR gains compared with two-dimensional constellations. Unfortunately, there exist only a few examples with equal set sizes.

Example 3.7 *We consider the Hurwitz constellation from Example 3.6. For* $M_4 = 50$*, we can construct the Lipschitz constellation according to [FS15a]. Figure 3.6 depicts both constellations, where the squared minimum Euclidean distance of the Lipschitz constellation is normalized to* $\text{dist}_\text{E} = 1$*, since the Hurwitz constellation has* $\text{dist}_\text{E} = 1$*, by design. In order to visualize four-dimensional constellation points in a three-dimensional plot, the values of the fourth dimension are indicated by different markers. Additionally, the boundaries of the constellation (i.e. of the respective modulo function) are shown, which are given as four-dimensional hypercubes depending on* λ *(cf. Section 3.2.1). They are indicated with colored vertice-connecting lines. Green or light gray colored and magenta or gray colored lines connect vertices of the Voronoi region, with corresponding points of the Voronoi cell (points of the Voronoi cell, before (3.43) is performed) with values of* $a_4 = 0.5$ *and* $a_4 = -0.5$*, respectively. Dashed lines indicate borders of the Voronoi region, with different values* a_4 *of the corresponding points of the Voronoi cell. As can be seen from Figure 3.6, the Hurwitz constellation in the right plot has a denser packing and the points seem to be equally distributed. Note that the Voronoi region of the Lipschitz constellation has a higher volume with unoccupied places, due to the fact that the construction of the Lipschitz constellation* $\mathcal{L}^{(0)}_{13+9\text{i}+8\text{j}+6\text{k}}$ *is based on the superset* $\mathcal{L}_{13+9\text{i}+8\text{j}+6\text{k}}$ *with* $N(13 + 9\text{i} + 8\text{j} + 6\text{k}) = 350$ *elements. The superset is separated into seven subsets of equal size, where* λ *and thus the Voronoi region does not change. More details can be found in [FS15a].*

Figure 3.7 depicts the two projections of l4_50, the constellation is normalized to a squared minimum Euclidean distance of $\delta^2 = 1$*. It can be seen, that the points in the projection have a distance of* 0.5*, which is the same as for a projection with Hurwitz integers. This comes due to the fact, that both constellations are based on the same unit cell. The difference between these two sets are the orientation, an offset, and, the outer shape. For our construction, the first two are defined with Hurwitz integers by design and the outer shape is defined by the modulo function and the choice of* λ*. Hence, when no algebraic structure is necessary, the CFM value can be increased by numerical optimization of those three properties.*

In Table 3.3, for $M_4 = 50$ *we have the CFM value 1.1765, whereas the sphere packing l4_50 has 1.3448 and the Lipschitz constellation has 0.8403. Hence, the sphere packing l4_50 achieves an asymptotic SNR gain of* $10\log_{10}(1.3448/1.1765) \approx 0.58$ *dB compared with the Hurwitz constellation, whereas the Hurwitz constellation achieves an asymptotic gain of* $10\log_{10}(1.1765/0.8403) \approx 1.44$ *dB compared with the Lipschitz constellation. Figure 3.8 provides simulation results for transmission over the AWGN channel with these constellations. The l4_50 sphere packing outperforms the other constellations over the whole SNR regime. For a SER of* 10^{-5}*, the Hurwitz constellation achieves a gain of about 0.7 dB compared with the Lipschitz constellation. The l4_50 sphere packing shows a gain of about 0.6 dB compared with the Hurwitz constellation. The four-dimensional constellations transmit 2.82* bits_2*. Figure 3.8 illustrates also the performance of 8 QAM (two 4 QAM constellations, where one is rotated by* $45°$ *and scaled by a factor* $(1+\sqrt{3})/\sqrt{2}$*) with 3* bits_2*. With respect to the energy per bit, the Lipschitz and the 8 QAM constellation achieve practically the same performance, where the Hurwitz constellation achieves a gain of 0.7 dB and the l4_50 sphere packing a gain of 1.3 dB in comparison with 8 QAM.*

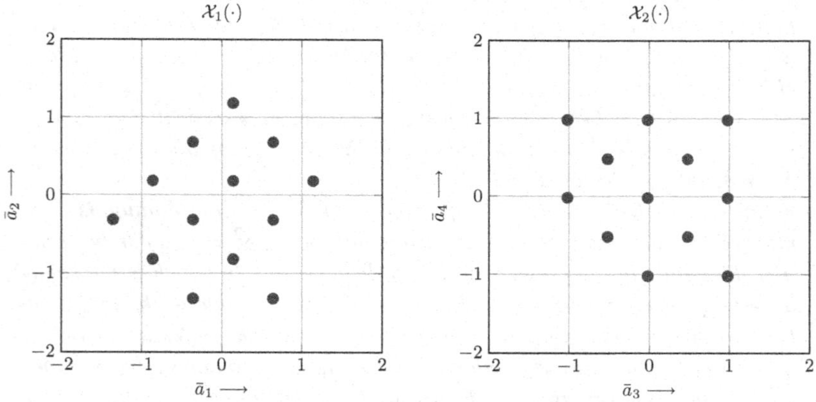

Figure 3.7 Example of the projections $\mathcal{X}_1$ and $\mathcal{X}_2$ for the sphere packing l4_50 from [Agr21]. The squared minimum Euclidean distance is normalized to $\delta^2 = 1$

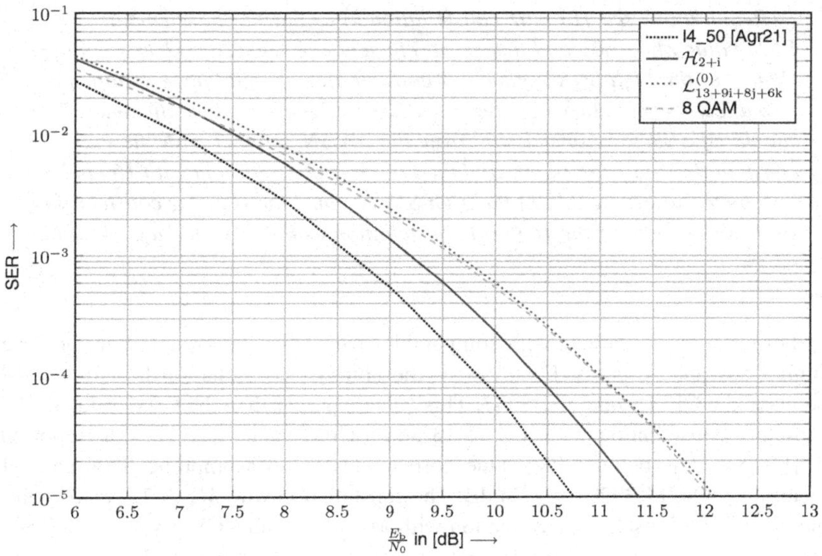

Figure 3.8 SER over the SNR per information bit for transmission over the AWGN channel for Hurwitz constellation $\mathcal{H}_{2+i}$ and Lipschitz constellation $\mathcal{L}^{(0)}_{13+9i+8j+6k}$ from [FS15a], and, the four-dimensional sphere packing 14_50 from [Agr21]. Each set includes $M_4 = 50$ points. Moreover, the figure depicts the simulation result for 8 QAM

Further comparisons with constellations from [Agr21] are presented in Table 3.4. This comparisons shows that the CFM values can be increased by numerical optimization. However, the numerical optimized constellations will not provide any algebraic structure and do not support a modulo reduction. On the other hand, a modulo function is required for techniques like Tomlinson-Harashima precoding [RCO+17] or advanced schemes like integer forcing [SF18].

Example 3.8 *We consider* $\lambda = 3 + 2i$ *with norm* $q = N(\lambda) = 13$. *Hence,* $\mathcal{L}_{3+2i}$ *has* $q^2 = 169$ *elements. Using* $\mathcal{L}_{3+2i}$ *as a signal constellation for transmission over the AWGN channel is equivalent to the transmission with Gaussian integers from the set* $\mathcal{G}_{3+2i}$. *Both constellations have a CFM value of* 0.4643 *and a spectral efficiency of* ≈ 3.7 bits_2.

Remarkably, the Hurwitz constellation $\mathcal{H}_{3+2\mathrm{i}}$ with 338 elements has a CFM value of 0.4602 *and a spectral efficiency of* $\log_2(2q^2)/2 \approx 4.2$ bits$_2$. *Hence, in the high-SNR regime, all constellations should have similar performance, but the Hurwitz-integer set has a higher spectral efficiency. The simulated SER for transmission over the AWGN channel with the Hurwitz constellation with 338 points and the Lipschitz constellation with 169 points is depicted in Figure 3.9. In the high-SNR regime, the Hurwitz constellation slightly outperforms the 16 QAM constellation with 4* bits$_2$ *and a CFM value of* 0.4.

For other constellations with comparable sizes, Hurwitz constellations achieve similar asymptotic gains. For $q = 3$, the Hurwitz constellation has $M_4 = 18$ points and a CFM value of 1.8947. This value is larger than the CFM value of the Lipschitz constellation with $M_4 = 13$ and a CFM value of 1.3929, as reported in [FS15a]. The better CFM value corresponds to an asymptotic SNR gain of $10 \log_{10}(1.8947/1.3929) \approx 1.34$ dB. The constellation with $M_4 = 72$ can be compared with the Lipschitz constellation with $M_4 = 73$ with a CFM value of 0.7399, where the improvement corresponds to an asymptotic SNR gain of 1.07 dB.

Table 3.4 CFM for the proposed Hurwitz-integer constellations and a comparison with optimized 4D constellations from [Agr21]

M_4	CFM proposed const.	const. from [Agr21]	CFM const. from [Agr21]
18	1.8947	c4_18	2.2326
50	1.1765	14_50	2.2326
72	0.9474	14_72	1.3448
98	0.8485	14_98	0.9565
200	0.5882	w4_200	0.6689

3.2.4 Suboptimal Signal Detection

Compared with two-dimensional constellations with the same spectral efficiency, the four-dimensional constellations have a disadvantage of a higher computational complexity for ML detection, since the search space increases from $2M_2$ for independent detection to $M_2^2 = M_4$. In this subsection, we present a suboptimal detection scheme for the four-dimensional constellations that achieves near-ML performance and significantly reduces the computational complexity.

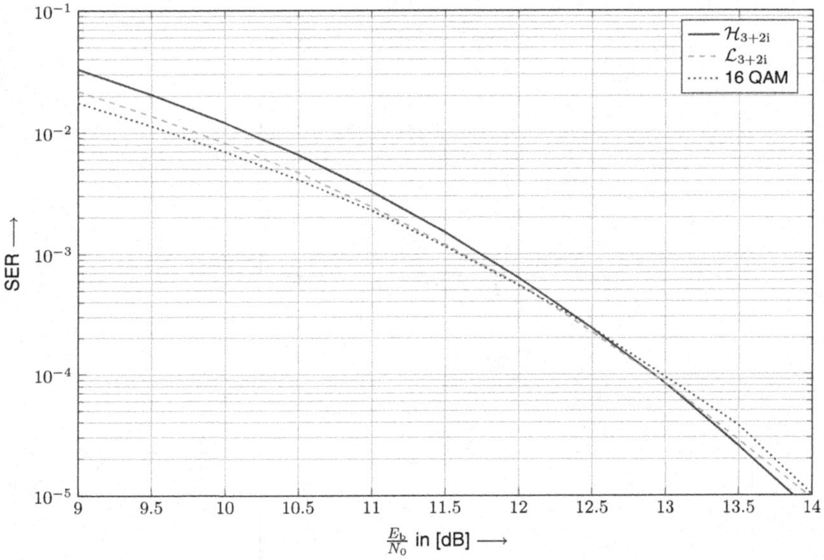

Figure 3.9 SER over the SNR per information bit for transmission over the AWGN channel for Hurwitz constellation $\mathcal{H}_{3+2i}$ with $M_4 = 338$ points, Lipschitz constellation $\mathcal{L}_{3+2i}$ with $M_4 = 169$ points, and 16 QAM

In order to reduce the search space for the symbol detection, we exploit the partitioning of the Hurwitz integers. This technique takes advantage of dependencies between elements of the projections $\mathcal{T}_1$ and $\mathcal{T}_2$. For instance, a Hurwitz integer has all elements from either the set of half-integers or the set of full-integers. Hence, if we select an integer from $\mathcal{T}_1$, this selections restricts the possible elements from $\mathcal{T}_2$ to the integer elements. Such dependencies can be used to simplify the detection procedure.

The transmission of an element from the Hurwitz constellation corresponds to the submission of two complex-valued symbols $x_1 \in \mathcal{T}_1$ and $x_2 \in \mathcal{T}_2$. We align the symbols to the symbol vector $\boldsymbol{x} = [x_1, x_2]$. First detecting the elements from one projection reduces the search space for detecting the signal point in the second projection. The subset of $\mathcal{T}_1$, which is determined by fixing an element $x \in \mathcal{T}_2$, is denoted by $\mathcal{T}_1(x)$. Similarly, the subset $\mathcal{T}_2(x) \subset \mathcal{T}_2$ is determined by an element $x \in \mathcal{T}_1$. An example with Hurwitz constellation $\mathcal{H}_{2+i}$ is given in Figure 3.10. In the uppermost plot projection $\mathcal{T}_1$, on the bottom left the conditional projection $\mathcal{T}_2(x_1)$,

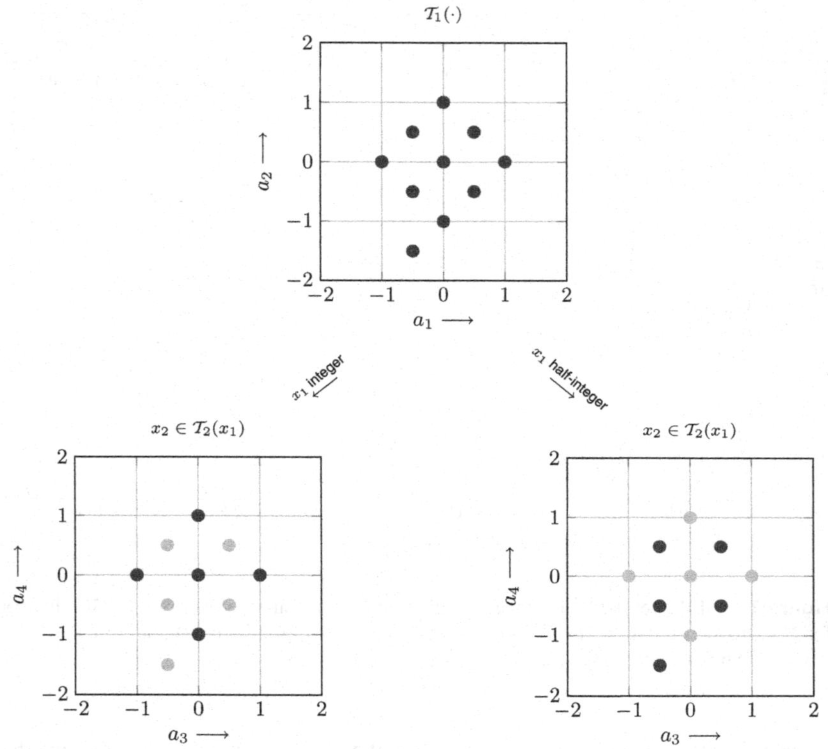

Figure 3.10 Example of the projection $\mathcal{T}_1$ for Hurwitz constellation $\mathcal{H}_{2+\mathrm{i}}$ with $M_4 = 50$ points. There are two conditional projections $\mathcal{T}_2(x_1)$, one for x_1 integer-valued and one for x_1 half-integer valued. The expurgated points are indicated in light gray

where x_1 is integer valued, and in the bottom right plot for the case x_1 is half-integer valued.

For a fixed value of x_1, we have $x_2 \in \mathcal{T}_2(x_1)$. Similarly, for a fixed value of x_2, we have $x_1 \in \mathcal{T}_1(x_2)$. We use this fact to determine two candidate vectors $\hat{\boldsymbol{x}}' = [\hat{x}_1', \hat{x}_2']$ and $\hat{\boldsymbol{x}}'' = [\hat{x}_1'', \hat{x}_2'']$ based on the received vector $\boldsymbol{y} = [y_1, y_2]$. The symbol $\hat{x}_1'$ is determined as the symbol from $\mathcal{T}_1$, which minimizes the Euclidean distance to the received value y_1. Next, $\hat{x}_2'$ is determined as the symbol from $\mathcal{T}_2(\hat{x}_1')$ based on y_2, i.e.

$$\hat{x}_1' = \underset{x \in \mathcal{T}_1}{\operatorname{argmin}} ||y_1 - x||^2 \text{ and} \tag{3.51}$$

$$\hat{x}_2' = \underset{x \in \mathcal{T}_2(\hat{x}_1')}{\operatorname{argmin}} ||y_2 - x||^2. \tag{3.52}$$

Similarly, $\hat{\boldsymbol{x}}''$ is obtained by first detecting $\hat{x}_2'' \in \mathcal{T}_2$ and $\hat{x}_1'' \in \mathcal{T}_1(\hat{x}_2'')$ afterward, i.e.

$$\hat{x}_2'' = \underset{x \in \mathcal{T}_2}{\operatorname{argmin}} ||y_2 - x||^2 \text{ and} \tag{3.53}$$

$$\hat{x}_1'' = \underset{x \in \mathcal{T}_1(\hat{x}_2'')}{\operatorname{argmin}} ||y_1 - x||^2. \tag{3.54}$$

Finally, the symbol vector is determined as

$$\hat{\boldsymbol{x}} = \underset{\boldsymbol{x} \in \{[\hat{x}_1', \hat{x}_2'], [\hat{x}_1'', \hat{x}_2'']\}}{\operatorname{argmin}} ||\boldsymbol{y} - \boldsymbol{x}||^2. \tag{3.55}$$

For large signal sets, the size of the projections is much smaller than the cardinality of the Hurwitz constellation. Consequently, the total number of comparisons and hence the computational complexity can be reduced. Note that the conditional projections can have different sizes. To demonstrate the complexity reduction, we present the average number of points in both projections M_2' in Table 3.5. Note that for the special case where λ is a Gaussian integer, all of the conditional projections have the same size. In Table 3.5, this holds for $q \in \{5, 10, 13, 17, 25, 29\}$. In this case, the number of comparisons does not depend on the detected symbols.

Figure 3.11 presents simulation results for the constellation with $M_4 = 968$ and a spectral efficiency of 4.96 bits_2. The performance of the suboptimal detection (denoted by (SD)) is practically identical to that of ML detection. Moreover, the figure provides the result for 32 QAM with 5 bits_2 for comparison.

On average, the detection according to (3.55) requires $2M_2'$ comparisons of complex-valued symbols, whereas ML detection requires $2M_4$ comparisons of complex-valued symbols. Hence, the average number of comparisons is reduced by the factor of M_4/M_2', which is in the range of 10 to 19 for the larger constellations in Table 3.5. The complexity of the proposed detection schemes is of order $\mathcal{O}(q)$, because the size of the projections is of order $\mathcal{O}(q)$. This becomes apparent in Figure 3.12, where the average number of comparisons per channel use is plotted for $q = 3, \ldots, 31$. The straight line is the regression line, which has a slope of 3.5. Compared with ML detection of two-dimensional symbols with comparable spectral efficiency, i.e. a constellation size of $M_2 \approx \sqrt{2q}$, the proposed detection is on average about 2.4 times more complex.

Table 3.5 Size of the Hurwitz-integer constellations and average number of symbols in the projections

q	M_4	M_2'	q	M_4	M_2'
3	18	11	18	648	77.5
5	50	15	19	722	71.2
6	72	22.8	21	882	75.5
7	98	24.9	22	968	80.7
9	162	31.5	25	1250	75
10	200	30	26	1352	92.3
11	242	40.3	27	1458	95.9
13	338	39	29	1682	87
15	450	54.2	31	1922	112.1
17	578	51			

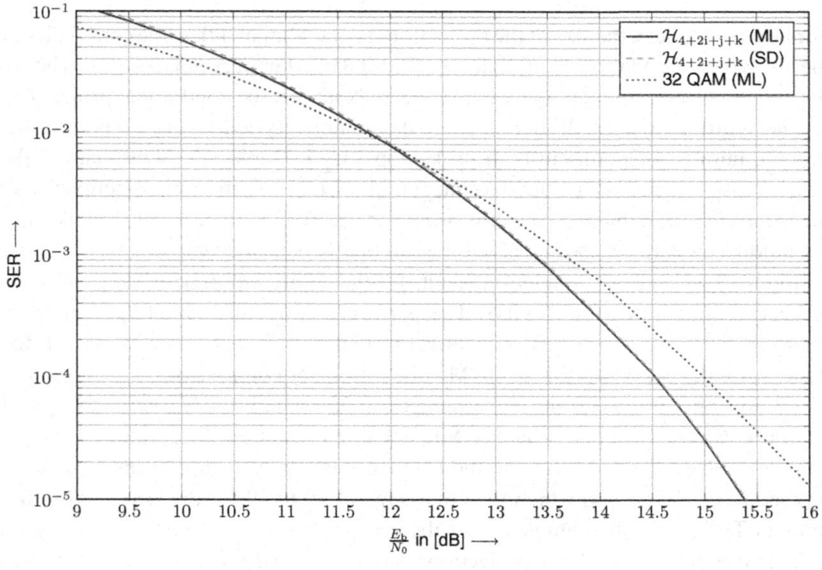

Figure 3.11 SER over the SNR per information bit for transmission over the AWGN channel for Hurwitz constellation $\mathcal{H}_{4+2i+j+k}$ with $M_4 = 968$ points and 32 QAM

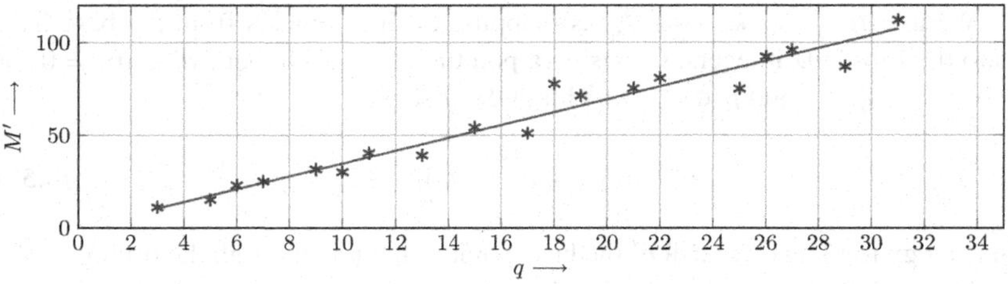

Figure 3.12 Average mean cardinality of the projection respectively the average number of comparisons per channel use over the order of the finite field is indicated with asterisks. The solid line is the regression line with slope 3.5

3.2.5 Set Partitioning

In the following, we present a second partitioning technique for Lipschitz and Hurwitz constellations that aims to find subsets with large squared Euclidean distance. In particular, we generalize the partitioning method presented in [FSS14] for Gaussian integers to the proposed Lipschitz and Hurwitz constellations. This set partitioning enables multilevel coding over Hurwitz constellations.

First, we consider the Lipschitz constellations, where $q = N(\lambda)$ is an arbitrary prime. In this case, the set $\mathbb{Z}_q^2$ is a two-dimensional Euclidean geometry over the base field $\mathbb{Z}_q$ [LC04]. The mapping $\mu_\lambda(\cdot)$ is operation-preserving with respect to addition and multiplication with elements from the base field. Therefore, $\mathcal{L}_\lambda$ is also a two-dimensional Euclidean geometry. In the following, we construct set partitions as lines (1-flats) in $\mathbb{Z}_q^2$ or $\mathcal{L}_\lambda$, respectively.

We use bold symbols to denote elements $\boldsymbol{a} \in \mathbb{Z}_q^2$, i.e. points in the Euclidean geometry. Let $\boldsymbol{a} \in \mathbb{Z}_q^2$ be a non-zero point. Then, the q points $\{\beta\boldsymbol{a}|\beta \in \mathbb{Z}_q\} \subset \mathbb{Z}_q^2$ and correspondingly the points $\{\mu_\lambda(\beta\boldsymbol{a})|\beta \in \mathbb{Z}_q\} \subset \mathcal{L}_\lambda$ form lines that pass through the origin (the zero-point). Let $\boldsymbol{a}_0$ and $\boldsymbol{a}$ be two points in $\mathbb{Z}_q^2$ such that $\beta\boldsymbol{a} + \boldsymbol{a}_0 \neq 0 \forall \beta \in \mathbb{Z}_q$. Then the q points $\{\mu_\lambda(\beta\boldsymbol{a} + \boldsymbol{a}_0)|\beta \in \mathbb{Z}_q\}$ form a line through the point $\boldsymbol{a}_0$.

The two lines $\beta\boldsymbol{a}$ and $\beta\boldsymbol{a} + \boldsymbol{a}_0$ are parallel lines, i.e. they have no common points. Consequently, the set $\{\mu_\lambda(\beta\boldsymbol{a} + \boldsymbol{a}_0)|\beta \in \mathbb{Z}_q\}$ is a coset of $\{\mu_\lambda(\beta\boldsymbol{a})|\beta \in \mathbb{Z}_q\}$. In this finite Euclidean geometry, there are $q - 1$ parallel lines for any line through the origin. This fact enables a partition of the set of Lipschitz integers $\mathcal{L}_\lambda$ into parallel lines.

We use $\boldsymbol{a}_k = \mu_\lambda(k), k \in \mathbb{Z}_q$ as mapping of the elements from the base field onto the Lipschitz integers. Choosing a point $\boldsymbol{a}' \in \mathbb{Z}_q^2$ such that $\beta\boldsymbol{a}' + \boldsymbol{a}_k \neq 0$ for $k = 1, \ldots, q-1$ and $\beta \in \mathbb{Z}_q$, we obtain the q subsets

$$\mathcal{L}_\lambda^{(k)} = \{\mu_\lambda(\beta\boldsymbol{a}' + \boldsymbol{a}_k) | \beta \in \mathbb{Z}_q\} \tag{3.56}$$

in analogy to (3.44). We call $\boldsymbol{a}'$ the base point of the line through the origin.

If $q = N(\lambda)$ is not a prime, then $\mathbb{Z}_q$ is a ring and $\mathbb{Z}_q^2$ is a module. In this case, (3.56) defines a coset if the real part and all imaginary parts of $\boldsymbol{a}'$ contain no zero divisors.

Similar to (3.44) and (3.45), we obtain the corresponding coset of half-integers by

$$\mathcal{V}_\lambda^{(k)} = \{\mu_\lambda(\boldsymbol{h} + \boldsymbol{a}'') : \boldsymbol{h} \in \mathcal{L}_\lambda^{(k)}\}, \tag{3.57}$$

where $\boldsymbol{a}''$ can be any element from the set of half-integers $\mathcal{V}_\lambda$. We call $\boldsymbol{a}''$ the coset leader. Finally, the subsets of Hurwitz integers are

$$\mathcal{H}_\lambda^{(k)} = \mathcal{L}_\lambda^{(k)} \cup \mathcal{V}_\lambda^{(k)}, \tag{3.58}$$

where $\mathcal{H}_\lambda^{(0)} \subset \mathcal{H}_\lambda$ is an additive subgroup of $\mathcal{H}_\lambda$ with $2q$ elements and $\mathcal{H}_\lambda^{(k)}, k = 1, \ldots, q-1$ are its cosets.

In the following, we estimate the minimum squared Euclidean distance δ^2 in the subsets. We use the following lemma on the modulo function.

Lemma 3.1 *For any Hurwitz integer a, we have*

$$||\mu_\lambda(a)||^2 \leq ||a||^2. \tag{3.59}$$

Proof. Let a_1, a_2, a_3, a_4 be the components of the quaternion $a\lambda^{-1} = a_1 + a_2\text{i} + a_3\text{j} + a_4\text{k}$. Then, with (3.43) we have

$$\begin{aligned}
\mu_\lambda(a)\lambda^{-1} &= a\lambda^{-1} - \lfloor a\lambda^{-1} \rceil \\
&= a_1 - \lfloor a_1 \rceil + (a_2 - \lfloor a_2 \rceil)\text{i} + \\
&\quad (a_3 - \lfloor a_3 \rceil)\text{j} + (a_4 - \lfloor a_4 \rceil)\text{k}.
\end{aligned}$$

Due to rounding, we have

$$|a_l - \lfloor a_l \rceil| \leq \frac{1}{2} \leq |a_l| \tag{3.60}$$

for any non-zero element a_l and $|a_l - \lfloor a_l \rceil| = 0$ for $a_l = 0$. Consequently, we obtain

$$||\mu_\lambda(z)\lambda^{-1}||^2 \leq ||z\lambda^{-1}||^2$$

and thus $||\mu(z)||^2 \leq ||z||^2$. □

Proposition 3.5 *Consider the set of Lipschitz integers $\mathcal{L}_\lambda$ and the q subsets $\mathcal{L}_\lambda^{(k)}$. The minimum squared Euclidean distance δ_L^2 of each subset satisfies*

$$\delta_L^2 \geq \min_{a \in \mathcal{L}_\lambda^{(0)} \backslash \{0\}} N(a). \tag{3.61}$$

Similarly, consider the set of Hurwitz integers $\mathcal{H}_\lambda$ and the q subsets $\mathcal{H}_\lambda^{(k)}$. The minimum squared Euclidean distance δ_H^2 of each subset satisfies

$$\delta_H^2 \geq \min_{a \in \mathcal{H}_\lambda^{(0)} \backslash \{0\}} N(a). \tag{3.62}$$

Proof. Let $a \neq b$ be two elements from the same subset $\mathcal{H}_\lambda^{(k)}$. The squared Euclidean distance is

$$\text{dist}_{\text{E}}(a, b) = ||a - b||^2.$$

With $c = \mu_\lambda(a - b)$ and Lemma 3.1, we have

$$\text{dist}_{\text{E}}(a, b) \geq ||\mu_\lambda(a - b)||^2 = N(c), \quad c \in \mathcal{H}_\lambda^{(0)} \setminus \{0\}$$

Hence, the lower bound holds for all cosets. The proof for the Lipschitz integers follows the same approach. □

Proposition 3.5 enables an efficient search for subsets with a high minimum squared Euclidean distance. For a given constellation $\mathcal{H}_\lambda$, we first construct all possible additive subgroups of Lipschitz integers (lines through the origin) and choose the set with the best minimum distance δ_L^2. Essentially, we can choose any non-zero element from $\mathbb{Z}_q^2$ as the base point $\boldsymbol{a}'$. However, there are at most $q + 1$ different lines through the origin. Accordingly, we have to investigate at most $q+1$ different sets. For each set, we have to calculate $q-1$ vector norms to find a base point that generates the subset $\mathcal{L}_\lambda^{(0)}$ with the best minimum distance. Hence, the complexity of this first search is of order $\mathcal{O}\left(q^2\right)$.

Next, we determine the subset of Hurwitz integers by choosing an element $\boldsymbol{a}''$ from the set of half-integers $\mathcal{V}_\lambda$. This set has q^2 elements. For each vector, we construct the set $\mathcal{V}_\lambda^{(0)}$ according to (3.57) and determine the minimum distance of the set $\mathcal{H}_\lambda^{(0)} = \mathcal{L}_\lambda^{(0)} \cup \mathcal{V}_\lambda^{(0)}$. This requires calculating $2q-1$ vector norms for all sets. Hence, the overall search complexity is of order $\mathcal{O}\left(q^3\right)$.

Table 3.6 presents search results for $q = 5, \ldots, 31$, where δ_L^2 is the minimum squared Euclidean distance in the subset of Lipschitz integers and δ_H^2 is the minimum squared Euclidean distance in the subset of Hurwitz integers, respectively. The column denoted by $\boldsymbol{a}'$ contains the base points to construct the subsets of Lipschitz integers. The coset leaders, which are needed to construct the subset of Hurwitz integers, are provided in the last column. For $q = 3$, the partitioning does not lead to larger distances in the subsets. For $q = 7$ and all values $q \geq 11$, the minimum squared Euclidean distance is improved in both partitions.

3.2.6 Multilevel Coding

Next, we discuss a construction of multilevel codes based on the partitioning given in Section 3.2.5 of the new Hurwitz constellations. At the end of this subsection, the potential of those Hurwitz signal constellations with multilevel-coded transmission via LDPC codes over the AWGN channel obtained by Monte-Carlo simulations is shown.

The system model is depicted in Figure 3.13, where m denotes the number of levels. The m data streams $\langle \boldsymbol{u}_0 \rangle, \langle \boldsymbol{u}_1 \rangle, \ldots, \langle \boldsymbol{u}_{m-1} \rangle$ are parsed into blocks $\boldsymbol{u}_l \in \mathcal{U}_l^{K_l}$, where K_l denotes the block length of $\boldsymbol{u}_l$, which is also the code dimension. $\mathcal{U}_l$ is the alphabet in the l-th level. Every information block $\boldsymbol{u}_l$ is individually encoded to the codeword $\boldsymbol{c}_l$ of length N_c by the corresponding encoder ENC$_l$. The (coded) modulation rate is given by

$$R_g = \sum_{l=0}^{m-1} \log_2\left(|\mathcal{U}_l|\right) R_l \tag{3.63}$$

with the level-code rate

$$R_l = \frac{K_l}{N_c}. \tag{3.64}$$

Table 3.6 Partitioning of Hurwitz-integer constellations (minimum squared Euclidean distance in subset of Lipschitz integers δ_L^2, base point $\boldsymbol{a}'$ for the partitioning of the Lipschitz integers, minimum squared Euclidean distance in subset of Hurwitz integers δ_H^2, coset leader $\boldsymbol{a}''$ for partitioning of the Hurwitz integers)

q	δ_L^2	$\mu_\lambda(\boldsymbol{a}')$	δ_H^2	$\mu_\lambda(\boldsymbol{a}'')$
5	2	$1+\mathrm{j}$	1	$\frac{1}{2}+\frac{1}{2}\mathrm{i}+\frac{1}{2}\mathrm{j}+\frac{1}{2}\mathrm{k}$
6	2	$-\mathrm{i}-\mathrm{j}+\mathrm{k}$	1	$\frac{1}{2}+\frac{1}{2}\mathrm{i}+\frac{1}{2}\mathrm{j}+\frac{1}{2}\mathrm{k}$
7	2	$1+\mathrm{j}$	2	$\frac{1}{2}-\frac{1}{2}\mathrm{i}-\frac{1}{2}\mathrm{j}-\frac{3}{2}\mathrm{k}$
9	2	$1+\mathrm{j}$	1	$\frac{1}{2}+\frac{1}{2}\mathrm{i}+\frac{1}{2}\mathrm{j}+\frac{1}{2}\mathrm{k}$
10	3	$1-\mathrm{j}+\mathrm{k}$	1	$\frac{1}{2}+\frac{1}{2}\mathrm{i}+\frac{1}{2}\mathrm{j}+\frac{1}{2}\mathrm{k}$
11	2	$1+\mathrm{j}$	2	$\frac{3}{2}+\frac{1}{2}\mathrm{i}-\frac{1}{2}\mathrm{j}+\frac{3}{2}\mathrm{k}$
13	3	$1+2\mathrm{j}$	3	$\frac{1}{2}+\frac{1}{2}\mathrm{i}+\frac{1}{2}\mathrm{j}+\frac{3}{2}\mathrm{k}$
15	3	$1+2\mathrm{j}$	3	$-\frac{1}{2}-\frac{3}{2}\mathrm{i}-\frac{1}{2}\mathrm{j}-\frac{3}{2}\mathrm{k}$
17	5	$1+2\mathrm{j}$	3	$\frac{1}{2}+\frac{1}{2}\mathrm{i}-\frac{1}{2}\mathrm{j}-\frac{3}{2}\mathrm{k}$
18	3	$1-\mathrm{i}-\mathrm{j}-2\mathrm{k}$	3	$\frac{1}{2}+\frac{1}{2}\mathrm{i}+\frac{3}{2}\mathrm{j}+\frac{1}{2}\mathrm{k}$
19	3	$1+2\mathrm{j}$	3	$-\frac{1}{2}+\frac{3}{2}\mathrm{i}+\frac{3}{2}\mathrm{j}-\frac{1}{2}\mathrm{k}$
21	4	$2-\mathrm{j}+2\mathrm{k}$	3	$\frac{1}{2}+\frac{1}{2}\mathrm{i}+\frac{3}{2}\mathrm{j}+\frac{1}{2}\mathrm{k}$
22	4	$-2+\mathrm{i}$	3	$\frac{3}{2}-\frac{1}{2}\mathrm{i}-\frac{1}{2}\mathrm{j}+\frac{5}{2}\mathrm{k}$
25	5	$1+2\mathrm{j}$	3	$\frac{1}{2}+\frac{1}{2}\mathrm{i}-\frac{7}{2}\mathrm{j}-\frac{1}{2}\mathrm{k}$
26	6	$1+3\mathrm{j}$	3	$\frac{1}{2}+\frac{1}{2}\mathrm{i}+\frac{3}{2}\mathrm{j}+\frac{1}{2}\mathrm{k}$
27	5	$1+3\mathrm{j}$	3	$\frac{1}{2}+\frac{1}{2}\mathrm{i}+\frac{3}{2}\mathrm{j}+\frac{1}{2}\mathrm{k}$
29	6	$1-2\mathrm{j}+2\mathrm{k}$	3	$\frac{1}{2}+\frac{1}{2}\mathrm{i}+\frac{3}{2}\mathrm{j}+\frac{1}{2}\mathrm{k}$
31	5	$1+2\mathrm{j}$	5	$-\frac{1}{2}+\frac{3}{2}\mathrm{i}-\frac{1}{2}\mathrm{j}-\frac{3}{2}\mathrm{k}$

The constellation mapper $\mathcal{M}$ aligns the codewords and maps the code symbols to points of the signal constellation, which results in the transmit vector $\boldsymbol{x}$.

For instance, consider a multilevel-coding scheme with two levels. The source is assumed to provide two non-binary data streams. Symbols are taken from the alphabets $\mathcal{U}_0 = \{0, \dots, q-1\}$ and $\mathcal{U}_1 = \{0, \dots, 2q-1\}$, since the considered partitioning technique separates the constellation into q subsets with $2q$ points each. The data streams $\langle \boldsymbol{u}_0 \rangle$, $\langle \boldsymbol{u}_1 \rangle$ are parsed into blocks $\boldsymbol{u}_0$, $\boldsymbol{u}_1$ of length K_l. After encoding, we have the codewords $\boldsymbol{c}_0$ and $\boldsymbol{c}_1$. The constellation mapper aligns the two codewords as

$$\boldsymbol{C} = \begin{bmatrix} \boldsymbol{c}_0 \\ \boldsymbol{c}_1 \end{bmatrix} = \begin{bmatrix} c_{0,0} & c_{0,1} & \cdots & c_{0,N_c-1} \\ c_{1,0} & c_{1,1} & \cdots & c_{1,N_c-1} \end{bmatrix}. \tag{3.65}$$

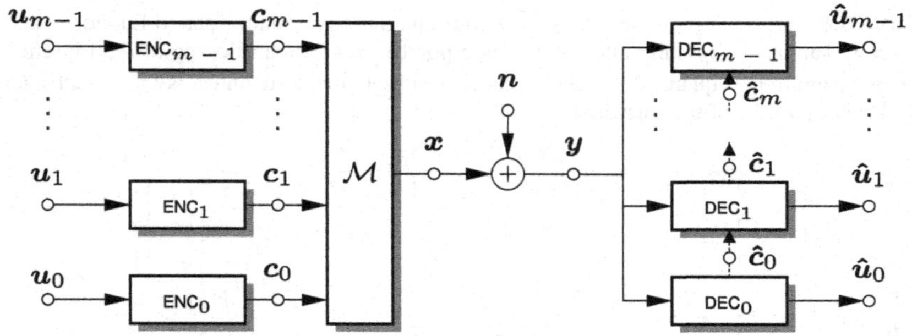

Figure 3.13 System model of multilevel-coded modulation for the proposed Hurwitz-integer constellations. Multilevel encoding and signal constellation mapper $\mathcal{M}$ on the left side, transmission over AWGN channel in the middle and multi-stage decoder on the right side

Every column is mapped to one point of the signal constellation as follows: the first level separates the constellation $\mathcal{H}_\lambda$ into q subsets $\mathcal{H}_\lambda^{(k)}$ with $2q$ symbols each, i.e. the code symbol of the first level $c_{0,i}$ defines a subset, and the code symbol of the second level $c_{1,i}$ defines the symbol within this subset, i.e. $x_i \in \mathcal{H}_\lambda^{(c_{0,i})} \subset \mathcal{H}_\lambda$ with $i = 1, \ldots, N_c - 1$.

After transmission, the noisy symbols in $\mathbf{y}$ are decoded successively, starting with level $l = 0$. First, likelihoods are calculated for the AWGN channel as

$$L_{j,i}^{(l)} = \sum_{h \in \mathcal{H}_\lambda^{(k)}} \exp\left(\frac{-|y_i - h|^2}{2\sigma_n^2}\right), \tag{3.66}$$

where $i = 0, \ldots, N_c - 1$ and $k = 0, \ldots, q - 1$. With these likelihoods, soft-input decoding is performed by DEC$_l$. This decoding is performed sequentially for all levels starting with level $l = 0$. This decoding results in the estimated codewords $\hat{\mathbf{c}}_l$ which are passed to the next level, where the code symbols are used to determine the Hurwitz subset for the next decoding step.

Figure 3.14 shows simulation results for transmission over the AWGN channel. The SNR is measured as the average energy per data symbol E_s over the noise power spectral density N_0 according to (2.7), to allow a comparison with the asymptotic SNR gain. The simulations consider the Hurwitz-integer constellation with $q = 19$ from Table 3.3. This constellation has $M_4 = 722$ symbols, which corresponds to a spectral efficiency of 4.75 bits$_2$. The black solid curve represents the SER for the uncoded transmission.

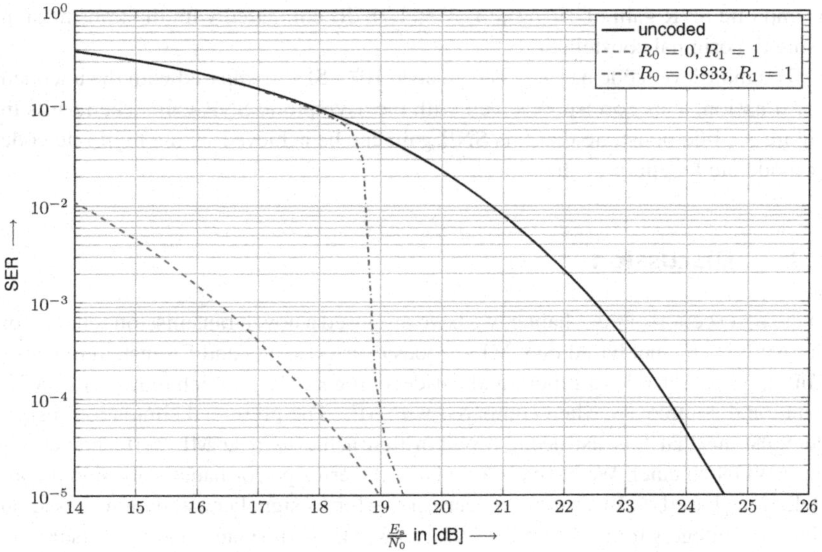

Figure 3.14 SER over the SNR in dB for transmission over the AWGN channel, multilevel-coded modulation with 4.39 bits_2 (dash-dotted line) as well as uncoded transmission of Hurwitz symbols with 2.62 bits_2 (dashed line) and 4.75 bits_2 (solid line)

For the multilevel coding, we consider a code with two levels, where only the first level is protected by a rate $R_0 = 0.833$ non-binary LDPC code of length $N_c = 1000$, while the second level remains uncoded. This results in a spectral efficiency of 4.39 bits_2. The non-binary LDPC code is taken from the subset of irregular repeat-accumulate codes [JKM00] to enable systematic encoding and it is created according to the semi-random construction described in [Ste19, Appendix A]. Non-binary belief-propagation decoding [DM98a, Ste19] (sum-product message passing) with up to 50 iterations is used. The simulation result corresponds to the red dash-dotted curve.

For comparison, the blue dashed curve presents results for the uncoded transmission with the subset $\mathcal{H}_\lambda^{(0)}$ according Table 3.6. This curve is a lower bound for the error rate in the second level, when we assume error-free decoding for the first level. Note that the subset has $M_2 = 38$ elements and a minimum squared Euclidean distance $\delta_H^2 = 3$. This corresponds to a spectral efficiency of 2.62 bits_2 and an asymptotic SNR gain of $10\log_{10}(\delta_H^2) \approx 4.78$ dB compared with the transmission using the original constellation.

The multilevel code has a 5 dB gain at $4 \cdot 10^{-5}$ SER compared with the uncoded transmission. This corresponds well with the asymptotic SNR gain. The results in Figure 3.14 demonstrate that this SNR gain can be achieved with a high-rate code of moderate length in the first level.

3.3 Discussion

In the first section of this chapter, we have proposed a novel transmission scheme for the AWGN channel based on OMEC codes over Gaussian and Eisenstein integers. This approach enables a suboptimal list-decoding strategy, which reduces the computational complexity when compared with ML. The presented simulation results demonstrate that this list-decoding algorithm achieves near-ML performance for the AWGN channel. We have observed that the error performance with signal constellations based on Eisenstein integers outperforms signal constellations based on Gaussian integers for mid-to-high SNR ratios. The performance gain for Eisenstein constellations comes due to a packing gain and a shaping gain. Gaussian integers are isomorphic to the integer lattice, whereas Eisenstein integers are isomorphic to the hexagonal lattice [CS99]. Both have the same squared minimum Euclidean distance.

The packing gain for Eisenstein integers comes due to the efficient positioning of points on the complex plane. It is known that the hexagonal lattice is the most efficient way to align points on a two-dimensional plane [CS99]. For low-SNR values, more errors occure due to the increased number of six nearest neighbors when compared with four nearest neighbors for Gaussian integers. A smaller contribution to the performance gain comes due to the shaping gain. Signal constellations based on Eisenstein integers are aligned in the densest way around the origin. The variance of the signal constellation is lower, which results in a lowered average energy in 2.7 and thus the curve is shifted in the direction of decreasing SNR values.

For higher dimensions, the question on the densest packing remains open. The most dense known packings can be found in the literature, e.g. in [CS99, CS03]. In four dimensions, Hurwitz integers are the densest known packing of points. In the second section of this chapter, we have presented a new construction method for four-dimensional signal constellations. These finite sets of Hurwitz integers have algebraic group properties and improved constellation figure of merit values compared with the four-dimensional Lipschitz-integer constellations from [FS15a] or the two-dimensional constellations from the first section. In particular, we have provided moduli and CFM values for all possible Hurwitz constellations with a size of up to 1922 elements.

In order to reduce the complexity for the symbol detection, a partitioning method was proposed that exploits dependencies between dimensions. This method achieves a near-ML performance with a reduced computational complexity. Simulation results for the AWGN channel demonstrate that the new Hurwitz constellations with suboptimal detection can outperform conventional two-dimensional constellations with ML detection for high-SNR ratios.

Furthermore, the partitioning method presented in [FSS14] for Gaussian integers was generalized to the proposed Hurwitz integer constellations. The subsets are additive subgroups and they can have much higher squared minimum Euclidean distances than the original constellations. This partitioning enables multilevel coding over Hurwitz constellations.

4 Spatial Modulation with Two-Dimensional Signal Constellations

In this chapter, we consider SM with one active transmit antenna per time instance. In the first section, we start with the signal detection at the receiver for the conventional SM transmission approach from Section 2.2.1 [MHAY06, MHS+08, JGS08, YDX+15]. We discuss three known suboptimal detection schemes and propose an efficient new approach. Based on the proposed detection scheme, we discuss properties of signal constellations and demonstrate that Eisenstein integers are well suited for SM. In the second section, we consider a coding scheme based on codes over Gaussian and Eisenstein integers that improves the performance of SPM.

Depending on the spectral efficiency, optimal-ML detection may not be feasable and thus suboptimal detection methods with reduced computational complexity are necessary. Most of the known suboptimal detection methods are based on two stages. In the first stage, the antenna that sends a non-zero signal is detected.

Once the antenna pattern is known, the sent symbol is detected. This procedure is appropriate for SM transmission scenarios. Note that for GMSM with multiple active transmit antennas, the search space for the second stage increases exponentially with the number of active transmit antennas and thus a comparison of all remaining possible vectors may be too computationally complex. A similar issue is apparent when SM with a signal constellation of high cardinality is used. We treat suboptimal detection methods for the second detection stage in the context of GMSM in Section 5.1.2. Those proposed methods can—if required—be adapted for SM transmission scenarios accordingly.

In Section 4.1, we focus on the first detection stage, analyze and compare the detection complexity of three known detection methods: SVD [WJS12b, PX13], MRC [NXQ11], and the GAM [LWL15]. The GAM is the best performing suboptimal detection method of those three, but it comes at the cost of the highest

D. B. Rohweder, *Signal Constellations with Algebraic Properties and their Application in Spatial Modulation Transmission Schemes*, Schriftenreihe der Institute für Systemdynamik (ISD) und optische Systeme (IOS), https://doi.org/10.1007/978-3-658-37114-2_4

computational complexity. We show that the computational complexity of this method can be significantly reduced without influencing the detection performance.

With two-stage detection, the overall error probability is bounded by the sum of the error probabilities in the first and second stage. We show that signal constellations based on the Eisenstein-integer lattice are suitable for SM transmission and they asymptotically balance the error probabilities for both stages. In contrast to the construction in Section 3.1.1, the signal constellations' cardinality can be independently chosen. The first section closes with simulation results.

In the second section, we consider SPM as a transmission technique from Section 2.2.4 [LSL+19]. In the original proposal of SPM, short repetition codes and permutation codes were used to construct a space-time code. We replace the repetition codes in SPM with codes that achieve a larger coding gain. We show the construction of a similar coding scheme that combines permutation codes with codes over Gaussian and Eisenstein integers. In contrast to repetition codes, codes over Gaussian and Eisenstein integers have good distance properties and achieve a significant coding gain for transmission over the AWGN channel [FGS13]. At the end of the section, simulation results show that this coding scheme outperforms the original proposal.

Parts of this chapter have already been published in: [3, 5, 6].

4.1 Spatial Modulation

In this section, we review known detection techniques to estimate the transmitted signal for SM with one active transmit antenna. The computational complexity is analyzed and it is shown how the Gaussian approximation method can be modified to reduce the computational complexity. Based on suboptimal detection techniques, considerations when designing signal constellations are discussed. At the end of this section, simulation results are shown.

4.1.1 Known Detection Methods

The overall-ML detector, which jointly estimates the antenna pattern and the transmitted symbol, is optimal in terms of error performance, although it requires the highest number of computations. The probability-density function (PDF) for a given received vector $\boldsymbol{y}$ depending on the channel vector $\boldsymbol{h}_i$ and symbol s reads

$$\mathrm{P}(\boldsymbol{y} \mid \boldsymbol{h}_i, s) = \frac{\exp\left(-(\boldsymbol{y} - \sqrt{\rho}\boldsymbol{h}_i s)^{\mathsf{H}} \boldsymbol{\Sigma}^{-1} (\boldsymbol{y} - \sqrt{\rho}\boldsymbol{h}_i s)\right)}{\pi^{N_{\mathrm{rx}}} \det(\boldsymbol{\Sigma})}, \tag{4.1}$$

where Σ denotes the covariance matrix. We replace Σ with the normalized noise covariance matrix $\boldsymbol{I}_{N_{\mathrm{rx}}}$ and can rewrite (4.1) as [LWL15, NXQ11]

$$\mathrm{P}(\boldsymbol{y} \mid \boldsymbol{h}_i, s) = \frac{1}{\pi^{N_{\mathrm{rx}}}} \exp\left(-||\boldsymbol{y} - \sqrt{\rho}\boldsymbol{h}_i s||^2\right). \tag{4.2}$$

With equiprobable use of antenna patterns, the overall-ML detector estimates the transmitted symbol vector as [JGS08]

$$(\hat{i}, \hat{s}) = \underset{i \in \{1,\ldots,N_{\mathrm{p}}\}, s \in \mathcal{T}}{\operatorname{argmin}} \; ||\boldsymbol{y} - \sqrt{\rho}\boldsymbol{h}_i s||^2, \tag{4.3}$$

where the minimization is performed over all possible transmitted symbol vectors. Similar to conventional wireless-MIMO systems with time-varying channels, channel estimation with SM is performed by the transmission of known signals, which are called pilots. It has been shown that SM is quite robust to channel estimation errors when compared with the Vertical-Bell Lab Layered Space-Time (V-BLAST) communication architecture [BAPP12, FAZ09, Fos96]. Various channel estimation schemes for SM, e.g. in [ADP14, SH12, WCDH14] et al., are known from the literature. In the following, we focus on the detection at the receiver and assume that it has perfect channel state information (CSI).

With ML detection, the computational complexity increases with order $\mathcal{O}\left(N_{\mathrm{p}} M_2\right)$ as more transmit antennas or larger modulation orders are employed. In order to reduce the detection complexity, multistage detection schemes were proposed where the active antenna is detected first and the transmitted symbol afterward [NXQ11, WJS12b, PX13]. These two-stage detection schemes reduce the complexity order from $\mathcal{O}\left(N_{\mathrm{p}} M_2\right)$ to $\mathcal{O}\left(N_{\mathrm{p}} + M_2\right)$.

We review suboptimal antenna detection approaches in the following subsections. For all considered suboptimal SM detection techniques, the transmitted symbol for a given estimated antenna pattern $\hat{i}$ is detected as

$$\hat{s} = \underset{s \in \mathcal{T}}{\operatorname{argmin}} \, ||\boldsymbol{y} - \sqrt{\rho}\boldsymbol{h}_{\hat{i}} s||^2. \tag{4.4}$$

Maximum Ratio Combiner

The MRC approach was introduced for SM by Meshleh et al. in [MHAY06]. In [NXQ11], a modified version was proposed to overcome required constraint channel conditions [JGS08, NXQ11]. With this modified approach, we first calculate

$$z_i = \boldsymbol{h}_i^{\mathsf{H}} \boldsymbol{y}. \tag{4.5}$$

Then, the index of the antenna pattern is estimated as

$$\hat{i} = \underset{i \in \{1,\ldots,N_\mathrm{p}\}}{\operatorname{argmax}} \frac{|z_i|}{||\boldsymbol{h}_i||}. \tag{4.6}$$

We apply the MRC detection rule for comparison reasons in terms of computational complexity and error performance in Sections 4.1.3 and 4.1.5, respectively. Furthermore, in Section 4.1.4 we use the MRC approach to derive signal constallation design rules for SM transmission scenarios.

Signal Vector-Based Detection

Next, we consider the SVD method [WJS12b, PX13] and demonstrate that this method is equivalent to the MRC approach. With SVD, the active antenna is detected as

$$\hat{i} = \underset{i \in \{1,\ldots,N_\mathrm{p}\}}{\operatorname{argmin}} \arccos\left(\frac{|\boldsymbol{h}_i^{\mathsf{H}} \boldsymbol{y}|}{||\boldsymbol{h}_i||\,||\boldsymbol{y}||}\right). \tag{4.7}$$

Due to the normalization, the argument in the inverse cosine is in the interval [0, 1] and the inverse cosine is strictly decreasing in this interval. Hence, SVD is equivalent to

$$\hat{i} = \underset{i \in \{1,\ldots,N_\mathrm{p}\}}{\operatorname{argmax}} \frac{|z_i|}{||\boldsymbol{h}_i||\,||\boldsymbol{y}||} = \underset{i \in \{1,\ldots,N_\mathrm{p}\}}{\operatorname{argmax}} \frac{|z_i|}{||\boldsymbol{h}_i||}, \tag{4.8}$$

where the last equation follows from the fact that the value $||\boldsymbol{y}||$ is the same for all $i \in \{1, \ldots, N_\mathrm{p}\}$. Hence, the SVD detection rule is equivalent to MRC (cf. (4.6) and (4.8)). It was demonstrated in [PX13] that SVD leads to a significant performance loss for high SNR values with QAM, whereas near-ML performance is optioned for PSK [BX16]. This fact motivates deriving design rules for signal constellations based on this detection rule.

Gaussian Approximation Method

The Gaussian approximation method [LWL15] is derived from the ML detection rule for the antenna pattern used

$$\hat{i} = \underset{i \in \{1,\ldots,N_\mathrm{p}\}}{\operatorname{argmax}} \mathrm{P}(\boldsymbol{y}|\boldsymbol{h}_i, s), \tag{4.9}$$

where $\mathrm{P}(\boldsymbol{y}|\boldsymbol{h}_i, s)$ is the PDF to receive $\boldsymbol{y}$ for given $\boldsymbol{h}_i$ and s, cf. (4.2). Every $\mathrm{P}(\boldsymbol{y}|\boldsymbol{h}_i, s)$ is a Gaussian distribution with expected value $\sqrt{\rho}\boldsymbol{h}_i s$. Remember that this method is used to estimate the antenna pattern i and the symbol s is estimated afterwards. Hence, the target is to calculate $\mathrm{P}(\boldsymbol{y}|\boldsymbol{h}_i)$ for every possible i, where each $\mathrm{P}(\boldsymbol{y}|\boldsymbol{h}_i)$ is the sum of M_2 Gaussian distributions. The idea of GAM is to approximate $\mathrm{P}(\boldsymbol{y}|\boldsymbol{h}_i)$ by a multivariate Gaussian distribution with zero-mean, since $\boldsymbol{y}$ is zero-mean and the influence of s is small, i.e. $\mathrm{P}(\boldsymbol{y}|\boldsymbol{h}_i, s) \approx \mathrm{P}(\boldsymbol{y}|\boldsymbol{h}_i)$, escpecially in the low-SNR range. We use (4.1) and replace the covariance matrix $\boldsymbol{\Sigma}$ by $\rho \boldsymbol{h}_i \boldsymbol{h}_i^{\mathsf{H}} + \boldsymbol{I}_{N_{\mathrm{rx}}}$. With an expected value of zero, the approximation reads

$$\mathrm{P}(\boldsymbol{y}|\boldsymbol{h}_i) \approx \frac{\exp(-\boldsymbol{y}^{\mathsf{H}}(\rho \boldsymbol{h}_i \boldsymbol{h}_i^{\mathsf{H}} + \boldsymbol{I}_{N_{\mathrm{rx}}})^{-1}\boldsymbol{y})}{\pi^{N_{\mathrm{rx}}} \det(\rho \boldsymbol{h}_i \boldsymbol{h}_i^{\mathsf{H}} + \boldsymbol{I}_{N_{\mathrm{rx}}})}. \tag{4.10}$$

This approximation leads to the GAM detection rule

$$\hat{\imath} = \operatorname*{argmax}_{i \in \{1,\ldots,N_{\mathrm{p}}\}} \frac{\exp\left(-\mathrm{y}^{\mathsf{H}}(\rho \boldsymbol{h}_i \boldsymbol{h}_i^{\mathsf{H}} + \boldsymbol{I}_{N_{\mathrm{rx}}})^{-1}\boldsymbol{y}\right)}{\det(\rho \boldsymbol{h}_i \boldsymbol{h}_i^{\mathsf{H}} + \boldsymbol{I}_{N_{\mathrm{rx}}})}. \tag{4.11}$$

4.1.2 Modified Gaussian Approximation Method

In the following, we simplify the GAM detection rule from (4.11) and define it as MGAM. Consider the covariance matrix

$$\boldsymbol{\Sigma} = \rho \boldsymbol{h}_i \boldsymbol{h}_i^{\mathsf{H}} + \boldsymbol{I}_{N_{\mathrm{rx}}}, \tag{4.12}$$

which is the sum of an invertible matrix $\boldsymbol{I}_{N_{\mathrm{rx}}}$ and the outer product $\rho \boldsymbol{h}_i \boldsymbol{h}_i^{\mathsf{H}}$. Applying the Sherman-Morrison formula (see, e.g. [GL12]), we obtain the inverse

$$\Sigma^{-1} = (\rho \boldsymbol{h}_i \boldsymbol{h}_i^{\mathsf{H}} + \boldsymbol{I}_{N_{\mathrm{rx}}})^{-1} = \boldsymbol{I}_{N_{\mathrm{rx}}} - \frac{\rho \boldsymbol{h}_i \boldsymbol{h}_i^{\mathsf{H}}}{1 + \rho \boldsymbol{h}_i^{\mathsf{H}} \boldsymbol{h}_i} = \boldsymbol{I}_{N_{\mathrm{rx}}} - \frac{\rho \boldsymbol{h}_i \boldsymbol{h}_i^{\mathsf{H}}}{1 + \rho ||\boldsymbol{h}_i||^2}, \tag{4.13}$$

where the inverse exists for $\rho ||\boldsymbol{h}_i||^2 > 0$. Taking the logarithm of $\mathrm{P}(\boldsymbol{y}|\boldsymbol{h}_i)$ and using (4.13) yields

$$\hat{\imath} = \operatorname*{argmax}_{i \in \{1,\ldots,N_{\mathrm{p}}\}} \frac{\rho \boldsymbol{y}^{\mathsf{H}} \boldsymbol{h}_i \boldsymbol{h}_i^{\mathsf{H}} \boldsymbol{y}}{1 + \rho ||\boldsymbol{h}_i||^2} - \log(\det(\rho \boldsymbol{h}_i \boldsymbol{h}_i^{\mathsf{H}} + \boldsymbol{I}_{N_{\mathrm{rx}}})). \tag{4.14}$$

Furthermore, we can simplify the rule to

$$\hat{i} = \underset{i \in \{1,\ldots,N_\mathrm{p}\}}{\operatorname{argmax}} \frac{\rho\,|z_i|^2}{1+\rho||\boldsymbol{h}_i||^2} - \log(1+\rho||\boldsymbol{h}_i||^2), \tag{4.15}$$

using

$$\det(\rho \boldsymbol{h}_i \boldsymbol{h}_i^\mathsf{H} + \boldsymbol{I}_{N_\mathrm{rx}}) = 1 + \operatorname{tr}\left(\rho \boldsymbol{h}_i \boldsymbol{h}_i^\mathsf{H}\right) = 1 + \rho||\boldsymbol{h}_i||^2. \tag{4.16}$$

For comparison reasons, again we use z_i. In contrast to the MRC rule in (4.6), the normalization is different. One additional term is present, which only depends on the SNR and the transmission path, i.e. the term is independent from the received signal.

4.1.3 Computational Complexity Analysis

Now, we consider the computational complexity of the different SM detection methods. For the suboptimal detection methods, some computations have to be performed for each received vector, whereas other computations are only required for each new channel instance. Hence, we estimate the computational complexity per received vector and per channel instance. We determine the computational complexity using multiply-accumulator (MAC) operations, where one MAC operation is the product of two (real- or complex-valued) numbers, added to a register, i.e. $a \leftarrow a + (b \cdot c)$. Note that argmin and argmax operations, e.g. in (4.14), are logical comparisons with the complexity $\mathcal{O}\left(N_\mathrm{p}\right)$ and they require no additional MAC operations.

First, we consider the computational complexity per received vector. For overall-ML detection according to (4.3), the computational complexity is dominated by the $M_2 N_\mathrm{p}$ matrix-vector products of the $(N_\mathrm{rx} \times N_\mathrm{tx})$ channel matrix $\boldsymbol{H}$ and the transmitted signal vector $\boldsymbol{x}$, which leads to $N_\mathrm{rx} N_\mathrm{tx}$ MAC operations per transmit vector and a total of $M_2 N_\mathrm{rx} N_\mathrm{tx}^2$ MAC operations per received vector.

For the suboptimal detection schemes MRC, SVD, and the proposed modified GAM (MGAM) method, the estimation of the antenna pattern requires N_p scalar products, each with N_rx MAC operations. This results in a total of $N_\mathrm{p} N_\mathrm{rx}$ MAC operations. The GAM detection is more computationally complex because a vector-matrix product is required. Assuming that we can solve the exponential function by a simple table lookup, the calculation of (4.11) needs at least N_rx^2 MAC operations per antenna pattern. For all suboptimal detection methods, using the estimated active antenna index, the symbol can be determined due to (4.4), which results in $M_2 N_\mathrm{rx}$ MAC operations.

Table 4.1 Estimation of the computational complexity in MAC operations for different detection methods. The necessary MAC operations per received vector and per channel instance are shown

Detection method	MAC op. per received vector	MAC op. per channel instance
overall-ML	$M_2 N_{rx} N_{tx} N_p$	0
MRC	$N_p N_{rx} + M_2 N_{rx}$	$N_p N_{rx}$
SVD	$N_p N_{rx} + M_2 N_{rx}$	$N_p N_{rx}$
GAM	$N_p N_{rx}^2 + M_2 N_{rx}$	$(2N_{rx}^3 + 2N_{rx})N_p$
MGAM	$N_p N_{rx} + M_2 N_{rx}$	$N_p N_{rx}$

Next, we consider the number of MAC operations per channel instance. For MRC and SVD, vector scaling is required according to (4.5) and (4.7), which results in $N_p N_{rx}$ operations. The GAM detection rule according to (4.11) is more complex due to the matrix inversion. Each matrix inversion needs N_{rx}^3 MAC operations and is required N_p times. Similarly, the computation of a determinant needs N_{rx}^3 operations per antenna pattern. Moreover, $2N_{rx}$ MAC operations are required to calculate $(\rho \boldsymbol{h}_i \boldsymbol{h}_i^{\mathsf{H}} + \boldsymbol{I}_{N_{rx}})$. This leads to a total of $(2N_{rx}^3 + 2N_{rx})N_p$ operations per channel instance. The proposed detection method simplifies GAM, because no matrix inverse and determinants are needed. Only the right term in (4.15) has to be calculated per channel instance, which leads to $N_p N_{rx}$ MAC operations.

Table 4.1 shows the estimated number of MAC operations. Figures 4.1 and 4.2 present numerical results depending on the alphabet size and antenna setup. We choose the highest possible number of antenna pattern, i.e. $N_p = N_{tx}$.

In Figure 4.1, it can be seen that for all detection techniques, the computational complexity for one received vector rises linearly with the cardinality of the signal constellation. For ML detection, the relation is one-to-one, since we have one multiplicative term only. The computational complexity with GAM and MGAM increases by a smaller factor and is always lower compared with ML detection. For higher cardinalities, GAM and MGAM asymptotically converge, where MGAM always remains lower.

Figure 4.2 depicts the computational complexity for one channel instance for an increasing number of receive antennas with the GAM method and the proposed modified version. As can be seen from the figure and Table 4.1, the main difference is that the number of MAC operations for the original proposal increases with order $\mathcal{O}\left(N_{rx}^3\right)$, whereas for the modified version it is $\mathcal{O}\left(N_{rx}\right)$.

4.1.4 Signal Constellation Design

Since suboptimal detection may lead to a performance loss, it is reasonable to design signal constellations specifically for a particular detection scheme to minimize the overall detection error probability. We believe that Eisenstein-integer constellations as proposed in [FS17] are suitable constellations for SM with suboptimal detection.

This is motivated by the following arguments. Consider the term in (4.6). For the actually-transmitted antenna index, we have

$$\frac{z_i}{||\boldsymbol{h}_i||} = \frac{\boldsymbol{h}_i^{\mathsf{H}} \boldsymbol{y}}{||\boldsymbol{h}_i||} = ||\boldsymbol{h}_i|| s + \frac{\boldsymbol{h}_i^{\mathsf{H}} \boldsymbol{n}}{||\boldsymbol{h}_i||} \tag{4.17}$$

and the triangle inequality

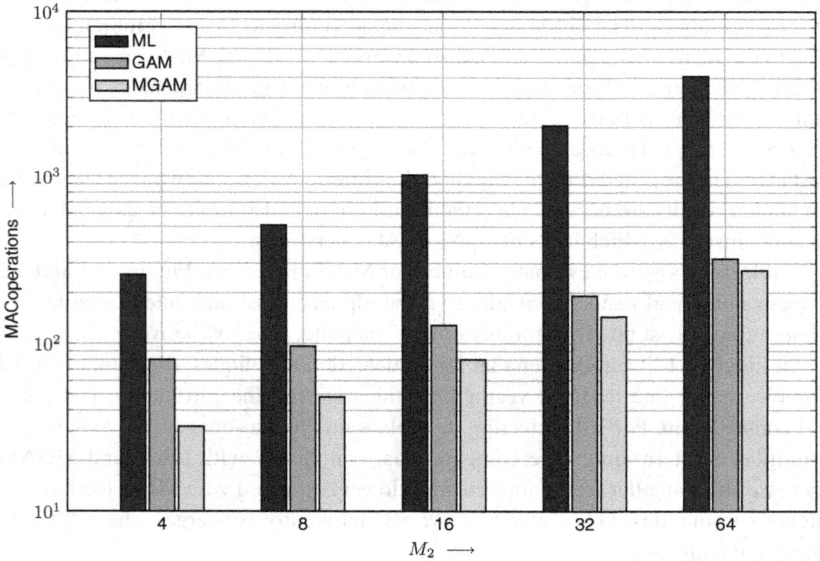

Figure 4.1 Number of MAC operations per received vector depending on the cardinality of the signal constellation M_2 with the setup of $N_{\text{tx}} = 4$ transmit and $N_{\text{rx}} = 4$ receive antennas. The number of antenna patterns is $N_{\text{p}} = 4$

$$|z_i| \leq ||\boldsymbol{h}_i||^2 |s| + \left|\boldsymbol{h}_i^{\mathrm{H}} \boldsymbol{n}\right|. \tag{4.18}$$

Hence, in order to maximize the right term in (4.18), we should use constellations with a high minimum energy E_{min} value. On the other hand, the detection of the transmitted symbol is asymptotically, i.e. at high SNR, determined by the minimum squared Euclidean distance in the signal constellation. Hence, in order to balance the error probabilities of the antenna selection and signal detection, we propose balancing the minimum energy and the minimum squared Euclidean distance in the signal constellation $\mathcal{T}$.

Now, consider the properties of the different signal constellations. PSK minimizes the error probability for the antenna pattern because $E_{\text{min}} = E_{\text{s}}$. However, PSK has a small minimum squared Euclidean distance for large values of M_2, i.e. $\delta^2 < E_{\text{min}}$ for $M_2 > 6$. Hence, with PSK, the error probability is asymptotically dominated by the detection of the transmitted symbol. On the other hand, for QAM we have $E_{\text{min}} = \frac{\delta^2}{2}$. Consequently, the antenna index is asymptotically more sensitive to detection errors than the selection of the symbol. Finally, the Eisenstein integers fulfill $E_{\text{min}} = \delta^2$ by design, which is a strong indicator of an asymptotically balanced error probability in both detection stages.

4.1.5 Simulation Results

In the following, we present results obtained from Monte-Carlo simulations over Rayleigh fading channels. Figure 4.3 depicts the two constellations that are used for simulations. We consider one transmission scheme with $N_{\text{rx}} = N_{\text{tx}} = 4$, where 10^7 symbol vectors were randomly generated per SNR value and transmitted over the channel. Detection was performed using the received noisy-signal samples, assuming perfect CSI at the receiver. Figures 4.4 and 4.5 present results for a spectral efficiency of 6 bpcu ($M_2 = 16$) and 7 bpcu ($M_2 = 32$), respectively.

For ML detection, Eisenstein constellations outperform QAM, whereas PSK shows a huge performance loss. MRC achieves ML performance for PSK, whereas there is a performance loss for the QAM and Eisenstein constellations with suboptimal detection, where the loss for the Eisenstein constellation is much smaller than for QAM. For 6 bpcu in Figure 4.4 with SNR values below 20 dB, the QAM constellation with MRC performs better than PSK. The Eisenstein constellations with MGAM outperform QAM with ML detection for SNR values below 15 dB. Considering the results for 7 bpcu in Figure 4.5, the tendencies are similar. With the Eisenstein constellations, the loss due to the suboptimal MGAM detection is less

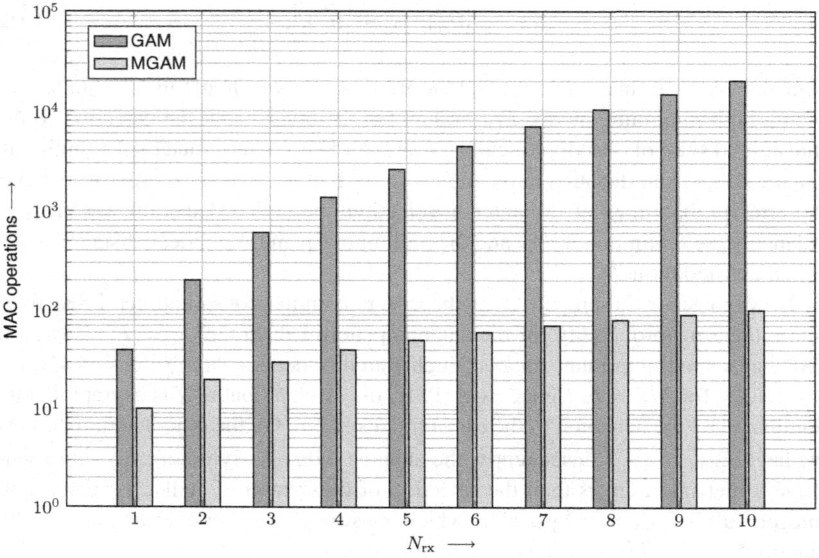

Figure 4.2 Number of MAC operations per channel instance depending on the number of receive antennas for $M_2 = 64$ constellation symbols and $N_{\text{tx}} = N_{\text{p}} = 10$

than 0.5 dB. Compared with MRC detection for QAM constellations, the proposed decoding method achieves a gain of more than 1 dB at high SNR values.

As mentioned in Section 4.1.4, the probability of a detection error depends on the minimum squared Euclidean distance δ^2 and the minimum energy E_{min}. Asymptotically, the squared Euclidean distance determines the probability of detecting an incorrect symbol within the constellation, whereas the minimum energy determines the probability of an erroneous decision with respect to the active antenna. For PSK, MRC achieves ML performance because the energy is equal for all signal points. This results in a low probability of an erroneous decision regarding the antenna. However, the error probability for the symbol is much higher due to the small distance δ^2. On the other hand, with QAM the suboptimal detection leads to a larger loss. Due to $E_{\text{min}} = \frac{\delta^2}{2}$, QAM is more sensitive to detection errors for the active antennas. Eisenstein constellations fulfill $E_{\text{min}} = \delta^2$, which asymptotically balances the error probabilities for the antenna detection and the symbol detection.

4.2 Space-Time Block-Coded Spatial Permutation Modulation

SPM is a recently-proposed transmission technique based on SM [LSL+19]. In contrast to SM, the channel is used multiple times to transmit the symbols of a codeword. SPM disperses the spatial symbol in time by using a permutation vector that indicates the active transmit antenna at successive time instances (cf. Section 2.2.4). The simplest space-time coding approach proposed in [LSL+19] combines short repetition codes and permutation codes. This SPM approach can be interpreted as a generalization of SM because the active transmit antenna index in SM is generalized to the permutation vector in SPM. Compared with SM, SPM achieves higher diversity and lower error rates. On the other hand, the coding approach is based on simple repetition codes, because the same symbol is transmitted multiple times using different antennas.

In this section, we replace the repetition codes in SPM with codes that achieve a larger coding gain. The SPM approach is based on very short repetition and permutation codes. We utilize the permutation codes proposed in [LSL+19] and extended in Table 2.1 to indicate the active transmit antenna at different time instances, but use codes over Gaussian and Eisenstein integers from Section 3.1.1 to encode the information. Since we have only one antenna that sends a non-zero signal with SPM, we can apply finite sets of Gaussian and Eisenstein integers as a signal constellation for the two-dimensional signal space.

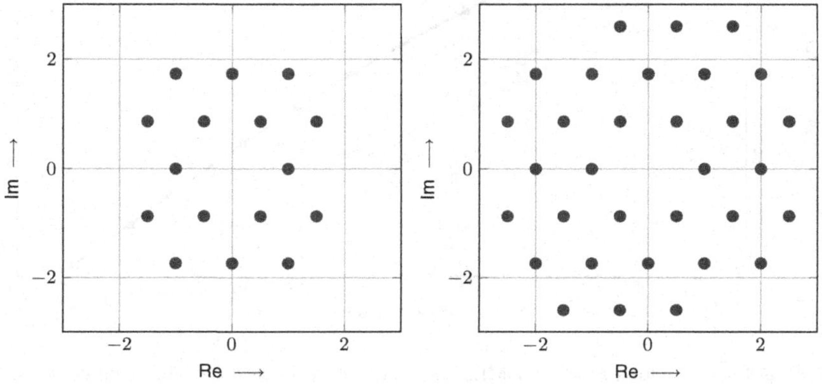

Figure 4.3 Eisenstein constellations with 16 and 32 elements

4.2.1 Space-Time Block-Code Construction

We combine SPM from Section 2.2.4 and OMEC codes from Section 3.1.1 to construct a space-time block (STB) code $\mathcal{C}_X$, where the codewords $\boldsymbol{X}$ are the signal matrices for SM transmission. In particular, the STB code $\mathcal{C}_X$ is obtained by combining a code $\mathcal{C}_{\mathcal{G}}(T, k, \mathrm{dist}_{\mathrm{M}})$ or $\mathcal{C}_{\mathcal{E}}(T, k, \mathrm{dist}_{\mathrm{M}})$ with a set of transmission patterns $\mathcal{P}\left(T, N_{\mathrm{tx}}, N_{\mathrm{p}}, \mathrm{dist}_{\mathrm{H}}\right)$ of the same length T. In particular, we choose the symbol vector $\boldsymbol{s} = [s_1, \ldots, s_T] \in \mathcal{C}_{\mathcal{G}}(T, k, \mathrm{dist}_{\mathrm{M}})$ or $\boldsymbol{s} \in \mathcal{C}_{\mathcal{E}}(T, k, \mathrm{dist}_{\mathrm{M}})$ as a codeword of the OMEC code. To map the codeword to a signal matrix $\boldsymbol{X}\left(\boldsymbol{p}_i, \boldsymbol{s}\right)$, we choose a transmission pattern $\boldsymbol{p}_i$ from the set $\mathcal{P}\left(T, N_{\mathrm{tx}}, N_{\mathrm{p}}, \mathrm{dist}_{\mathrm{H}}\right)$ in Table 2.1 randomly. In contrast to the original SPM approach, we do not use the choice of the transmission pattern to transmit information. However, the transmission pattern is required for the mapping from codeword to signal matrices. In principle, we could use a single transmission pattern for this mapping. In order to achieve a high transmit diversity

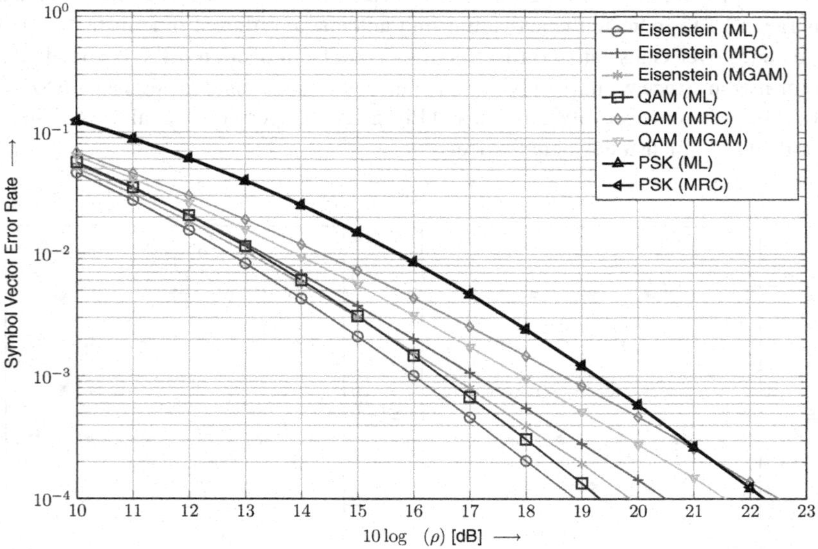

Figure 4.4 Simulation results for ML, MRC and MGAM detection. The setup comprises $N_{\mathrm{tx}} = N_{\mathrm{rx}} = 4$ transmit and receive antennas. The spectral efficiency is 6 bpcu, where 2 bpcu are mapped to the antenna pattern and the remaining bits are mapped to the symbol

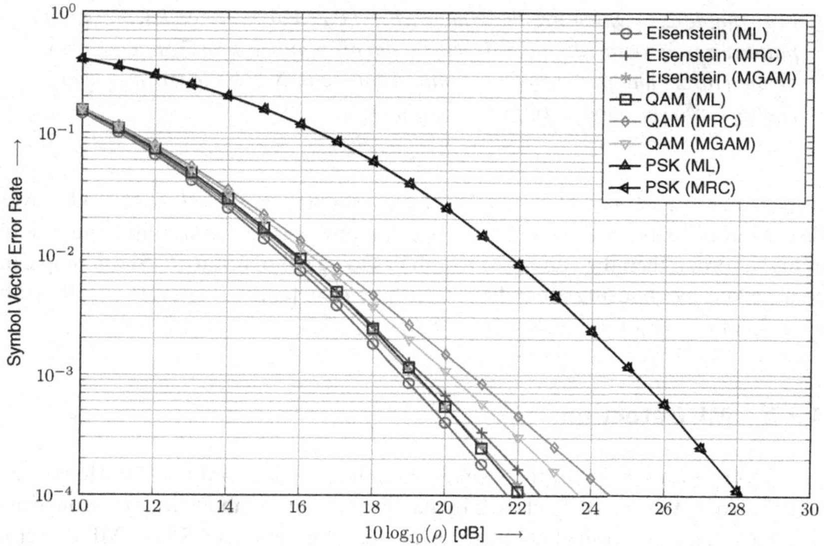

Figure 4.5 Simulation results for ML, MRC and MGAM detection. The setup uses $N_{\text{tx}} = N_{\text{rx}} = 4$ transmit and receive antennas. The spectral efficiency is 7 bpcu, where 2 bpcu are mapped to the antenna pattern and the remaining bits are mapped to the symbol

and equal usage of antennas, we choose a transmission pattern randomly for each codeword. This procedure results in a set of p^k signal matrices that form the STB code $\mathcal{C}_\mathcal{X}$. We demonstrate this construction for Gaussian integers in Example 4.1.

Example 4.1 *We construct an STB code* $\mathcal{C}_\mathcal{X}$ *for* $T = 3$ *and* $N_{\text{tx}} = 4$*. We choose the field size* $p = 13$ *to ensure* $T = \frac{(p-1)}{4} = 3$*. Hence, all transmitted symbols are from the finite Gaussian integer field* $\mathcal{G}_{13}$*. The corresponding OMEC code* $\mathcal{C}_\mathcal{G}(3, 2, 3)$ *has* $p^k = 13^2 = 169$ *codewords. For a binary mapping between information source and transmitter, we only consider* $2^7 = 128$ *codewords. For the set of transmission patterns, we select* $\mathcal{P}(3, 4, 24, 1)$*. Since no information bits are directly mapped to the transmission patterns, we choose* $\text{dist}_\text{H} = 1$ *and the largest possible value of* N_p *to ensure high* *transmit diversity. We construct the signal matrices $\boldsymbol{X}(\boldsymbol{p}, \boldsymbol{s})$ as described in Example 2.2, which are the codewords of $\mathcal{C}_X$. For every codeword $\boldsymbol{s}$, one pattern $\boldsymbol{p}$ is associated randomly with equal probability from $\mathcal{P}$, since we have defined direct mapping from information source to $\boldsymbol{X}$. The spectral efficiency is* $\lfloor \log_2(p^k) \rfloor = \lfloor \log_2(169) \rfloor = 7$ bpsq.

Note that if $T < \frac{p-1}{4}$ for codes over Gaussian integers, $T < \frac{p-1}{6}$ for codes over Eisenstein integers, or to increase the spectral efficiency by choosing a finite field of higher order, then the codeword length has to be shortened to T. Shortening can be achieved by choosing all codewords where $\boldsymbol{c} = [c_0, \ldots, c_{T-1}, 0, \ldots, 0]$ with $c_j \in \mathcal{G}_p$ or $c_j \in \mathcal{E}_p$ for which $\boldsymbol{P}\boldsymbol{c}^\mathsf{T} = 0$.

4.2.2 ML Detection

The information symbol of the source determines a codeword $\boldsymbol{s} \in \mathcal{C}_X$. Hence, the signal matrix $\boldsymbol{X}$ is uniquely chosen by the information symbol. After transmission, ML detection is performed on the receiver side. With perfect CSI, the ML detector estimates the transmitted symbols and the transmission pattern as

$$(\hat{\boldsymbol{p}}_i, \hat{\boldsymbol{s}}) = \underset{\substack{i \in \{1,\ldots,N_\mathrm{p}\}, \\ [s_1,\ldots,s_T] \in \mathcal{C}}}{\operatorname{argmin}} \sum_{\iota=1}^{T} ||\boldsymbol{y}_\iota - \sqrt{\rho}\boldsymbol{h}_{p_{i,\iota}} s_\iota||^2, \tag{4.19}$$

where the minimization is performed over all possible transmitted symbol vector sequences or codewords, respectively. $\boldsymbol{y}_\iota$ denotes the ιth column of the $N_\mathrm{rx} \times T$ received matrix $\boldsymbol{Y}$ and $\boldsymbol{h}_{p_{i,\iota}}$ is the $p_{i,\iota}^{\text{th}}$ column of the channel matrix $\boldsymbol{H}$. For the OMEC codes, the choise of the transmission pattern is random. In this case, due to the predifined mapping between transmission pattern and symbol vector or codeword, only the predefined transmission pattern is considered, i.e. every symbol vector or codeword inherently defines the assigned transmission pattern. The detection rule is given by

$$(\hat{\boldsymbol{p}}_i, \hat{\boldsymbol{s}}) = \underset{\substack{f: \mathcal{C} \to \mathcal{P},\, \boldsymbol{s} \mapsto i, \\ [s_1,\ldots,s_T] \in \mathcal{C}}}{\operatorname{argmin}} \sum_{\iota=1}^{T} ||\boldsymbol{y}_\iota - \sqrt{\rho}\boldsymbol{h}_{p_{i,\iota}} s_\iota||^2. \tag{4.20}$$

4.2.3 Simulation Results

In this section, we present results of Monte-Carlo simulations, which were obtained over Rayleigh fading channels. We consider a single transmission scheme with $N_{tx} = 4$ and $N_{rx} = 2$. In the simulations, 10^6 data symbols were randomly generated per SNR value and transmitted over the channel. ML detection according to (4.20) was performed using the noisy signal samples received, assuming perfect CSI at the receiver.

Figures 4.6 and 4.7 show simulation results for 7 bpsq with codes over Gaussian integers and 8 bpsq with codes over Eisenstein integers. All space-time codes have a length of $T = 3$. The curves labeled by QAM and PSK correspond to the original SPM scheme, where the repetition code $\mathcal{C}_{\text{Rep}}(3, 1, 3)$ is used and the QAM or PSK symbol is repeated T times, respectively. In this case, three or four information bits are mapped to the M_2-ary signal constellation as well as four information bits to select a transmission pattern from $\mathcal{P}(4, 4, 16, 2)$. In Figure 4.6, the SPM scheme with 8 PSK outperforms the transmission with 8 QAM. This results from the fact

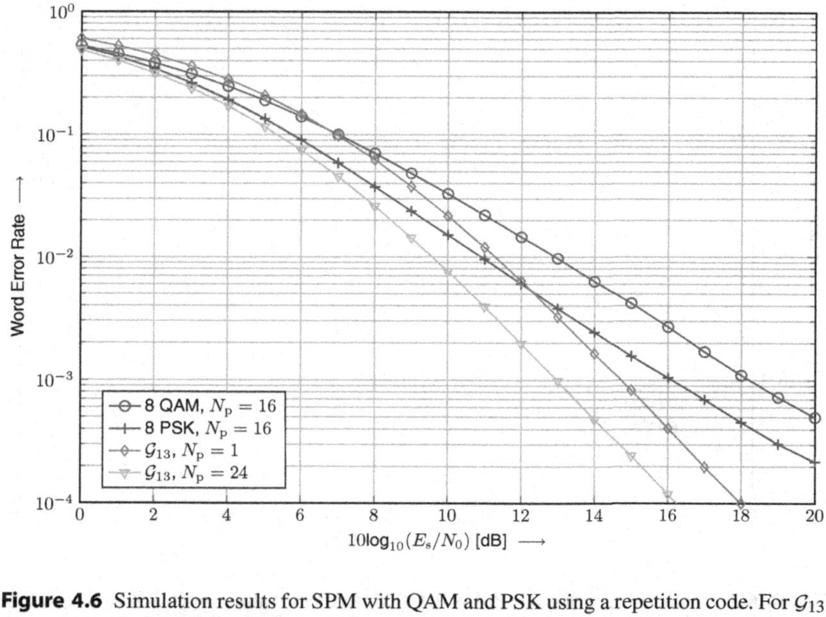

Figure 4.6 Simulation results for SPM with QAM and PSK using a repetition code. For $\mathcal{G}_{13}$, the code $\mathcal{C}_{\mathcal{G}}(3, 2, 3)$ from Example 4.1 is used. All codes have $T = 3$. The spectral efficiency is 7 bpsq

that all PSK symbols have the same energy, which is well suited for detecting the transmission pattern. The varying magnitudes of the QAM symbols lead to a higher error probability of detecting the transmission patterns, which impairs the overall detection performance. In Figure 4.7, similar behavior can be observed. The SPM scheme with 16 PSK outperforms the transmission with 16 QAM in the mid-to-high SNR regime. This comes due to the lowered squared Euclidean distance for 16 PSK, where the negative influence on the performance vanishes in the high-SNR regime.

In Figure 4.6, the two $\mathcal{G}_{13}$-labeled curves correspond to the proposed coding scheme according to Example 4.1 with Gaussian integers from the set $\mathcal{G}_{13}$. Using $N_\mathrm{p} = 24$ results in the highest transmit diversity and shows the best performance over the complete SNR range. Even for $N_\mathrm{p} = 1$, the OMEC code shows better performance for high-SNR values compared with SPM based on QAM and PSK. Note that with $N_\mathrm{p} = 1$, only one transmission pattern is used, which essentially reduces the system to $N_\mathrm{tx} = 3$ for $T = 3$ because one antenna is not selected.

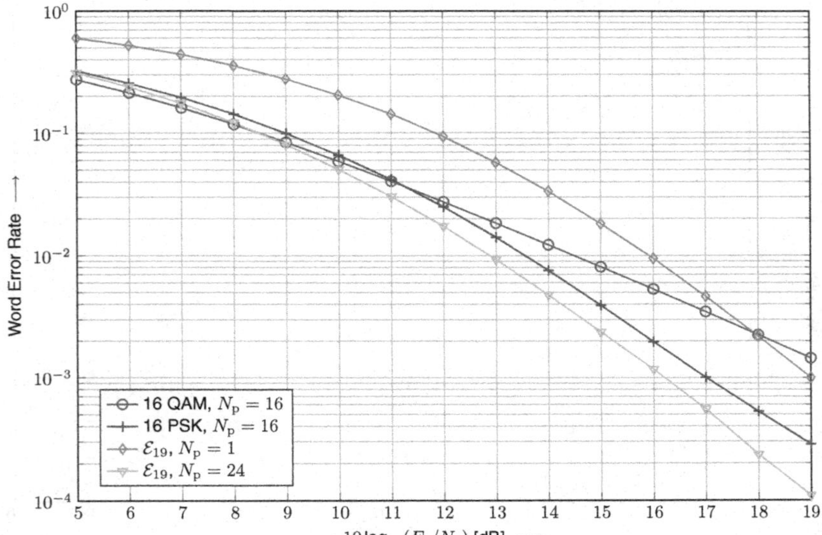

Figure 4.7 Simulation results for SPM with QAM and PSK using a repetition code. For $\mathcal{E}_{19}$, the code $\mathcal{C}_{\mathcal{E}}$ (3, 2, 3) is used. All codes have $T = 3$. The spectral efficiency is 8 bpsq

In Figure 4.7, the two $\mathcal{E}_{19}$-labeled curves correspond to the proposed coding scheme for codes over Eisenstein integers with symbols from the set $\mathcal{E}_{19}$. The results are similar to the 7 bpcu case with codes over Gaussian integers, where the same behavior shifted to a higher SNR regime can be observed.

Figure 4.8 depicts simulation results for a higher spectral efficiency with $T = 4$ and 12 bpsq. The SPM scheme with 256 QAM uses the repetition code $\mathcal{C}_{\text{Rep}}(4, 1, 4)$ and $\mathcal{G}_{17}$ uses the code $\mathcal{C}_{\mathcal{G}}(4, 3, 3)$. Furthermore, with 16 QAM we have the same spectral efficiency by concatination of two codewords of the repetition code $\mathcal{C}_{\text{Rep}}(2, 1, 2)$. As can be seen from the figure, $\mathcal{G}_{17}$ with largest possible value of $N_{\text{P}_{\max}}$ gives the best overall performance. For high SNR values > 25 dB, the SPM schemes with 16 QAM and 256 QAM have a similar performance, because the errors are dominated by misdetection of the transmission patterns. Both SPM schemes are based on $\mathcal{P}(4, 4, 16, 2)$ with the minimum Hamming distance $\delta^2 = 2$.

4.3 Discussion

This chapter provides an overview of transmission techniques with spatial modulation for one active transmit antenna. Known suboptimal detection methods to estimate the active transmit antenna were discussed. These methods reduce the computational complexity when compared with overall-ML detection. The Gaussian approximation method shows the best error performance but requires the highest number of computations. We have modified this method such that the computational complexity is reduced without influencing the error performance. Further reduction of the computational complexity can be achieved due to suboptimal symbol detection, which becomes more relevant with multiple-active transmit antennas and high modulation orders. Hence, this topic is treated in Section 5.1.2 for generalized multistream spatial modulation.

The signal constellation has a major impact on the system performance. When a two-stage detection technique is used, erroneous estimations can happen in the first as well as the second stage. Since the second stage is based on the result of the first stage, an incorrect estimation in the second stage is highly probable if the first stage provides an incorrect estimation. Based on suboptimal detection methods, considerations have been derived for designing signal constellations. Signal constellations with symbols drawn from the Eisenstein-integer lattice regard those considerations and show a lower error probability when compared with conventional signal constellations.

Furthermore, we have proposed a space-time coding scheme for SM, where only a single antenna is active per time instance. The code construction combines trans-

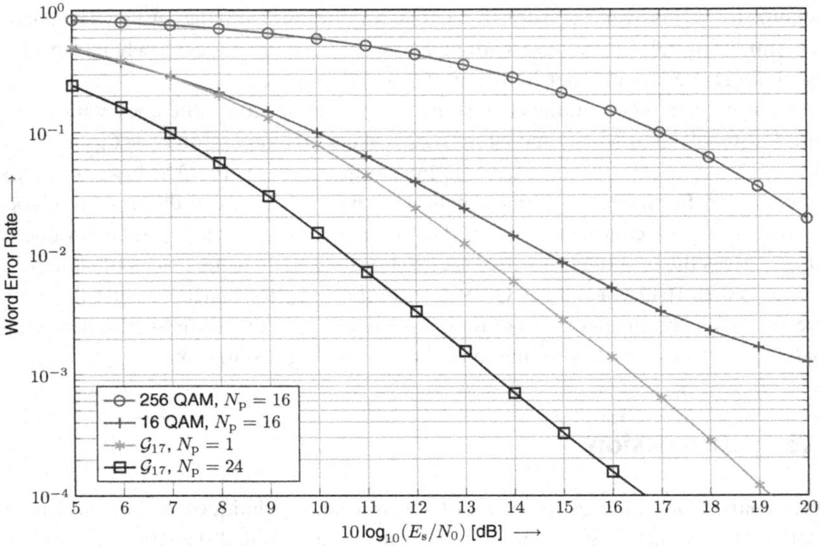

Figure 4.8 Simulation results for SPM with 12 bpsq. For $\mathcal{G}_{17}$, the code $\mathcal{C}_{\mathcal{G}}$ (4, 3, 3) is used. The 256 QAM SPM transmission scheme uses the repetition code $\mathcal{C}_{\text{Rep}}(4, 1, 4)$. For 16 QAM we use two concatinated codewords from the repetition code $\mathcal{C}_{\text{Rep}}(2, 1, 2)$

mission patterns from SPM with codes over Gaussian and Eisenstein integers. In the original proposal of SPM, the space-time codes constructed for a single active antenna are based on simple repetition codes. The proposed approach is motivated by the fact that repetition codes do not achieve a coding gain for transmission over the AWGN channel. On the other hand, short codes over Gaussian and Eisenstein integers have good distance properties and achieve a significant coding gain [FGS13]. In particular, the considered OMEC codes of maximum length are perfect codes with respect to the Mannheim distance [Hub94b, Hub94a, MBG07]. The presented simulation results demonstrate that the space-time codes based on OMEC codes outperform SPM with repetition codes.

The simulation results for SPM are based on ML decoding. However, the complexity of ML decoding increases exponentially with the code dimension. Hence, suboptimal decoding strategies would be desirable.

Spatial Modulation with Multidimensional Signal Constellations

5

The previous chapter has focused on two-dimensional signal constellations with SM, i.e. suitable signal constellations when only one transmit antenna sends a non-zero signal per time instance. In this chapter, we extend some of those approaches on constellation design, set partitioning, and detection methods to the case of more than one active transmit antenna, i.e. the application of multidimensional signal constellations with the GMSM transmission technique [YSMH10, WJS12a] from Section 2.2.3.

Parts of this chapter have already been published in: [2, 7, 8].

5.1 Generalized Suboptimal Detection

The overall-ML detector, which jointly estimates the antenna pattern and the transmitted symbols, is optimal in terms of error performance, although it requires the highest number of computations. The PDF for a given received vector $\boldsymbol{y}$ depending on the submatrix $\boldsymbol{H}_i$ of the channel matrix and symbol vector $\boldsymbol{s}$ reads

$$\mathrm{P}(\boldsymbol{y} \mid \boldsymbol{H}_i, \boldsymbol{s}) = \frac{\exp\left(-\left(\boldsymbol{y} - \sqrt{\rho}\boldsymbol{H}_i\boldsymbol{s}\right)^{\mathsf{H}} \boldsymbol{\Sigma}^{-1} \left(\boldsymbol{y} - \sqrt{\rho}\boldsymbol{H}_i\boldsymbol{s}\right)\right)}{\pi^{N_{\mathrm{rx}}} \det(\boldsymbol{\Sigma})}, \tag{5.1}$$

where $\boldsymbol{\Sigma}$ denotes the noise covariance matrix. We replace $\boldsymbol{\Sigma}$ with the normalized white noise covariance matrix $\boldsymbol{I}_{N_{\mathrm{rx}}}$ and can rewrite (5.1) as

$$\mathrm{P}(\boldsymbol{y} \mid \boldsymbol{H}_i, \boldsymbol{s}) = \frac{1}{\pi^{N_{\mathrm{rx}}}} \exp\left(-||\boldsymbol{y} - \sqrt{\rho}\boldsymbol{H}_i\boldsymbol{s}||^2\right). \tag{5.2}$$

D. B. Rohweder, *Signal Constellations with Algebraic Properties and their Application in Spatial Modulation Transmission Schemes*, Schriftenreihe der Institute für Systemdynamik (ISD) und optische Systeme (IOS), https://doi.org/10.1007/978-3-658-37114-2_5

With equiprobable use of antenna patterns, the overall-ML detector estimates the transmitted symbol vector as

$$(\hat{\imath}, \hat{s}) = \underset{i \in \{1,\dots,N_\mathrm{p}\}, s \in \mathcal{X}}{\operatorname{argmin}} \quad ||\boldsymbol{y} - \sqrt{\rho}\boldsymbol{H}_i \boldsymbol{s}||^2, \tag{5.3}$$

where the minimization is performed over all possible transmitted symbol vectors.

5.1.1 MGAM for GMSM

In Section 4.1.1, we have reviewed GAM for SM. In the following, we discuss this method for GMSM, which is based on the same idea. The GAM for GMSM is derived from the ML detection rule (5.2) for the antenna pattern as [LWL15]

$$\hat{\imath} = \underset{i \in \{1,\dots,N_\mathrm{p}\}}{\operatorname{argmax}} \ \mathrm{P}(\boldsymbol{y}|\boldsymbol{H}_i, \boldsymbol{s}), \tag{5.4}$$

where $\mathrm{P}(\boldsymbol{y}|\boldsymbol{H}_i, \boldsymbol{s})$ is the conditional PDF to receive $\boldsymbol{y}$ for a given channel submatrix $\boldsymbol{H}_i$ and $\boldsymbol{s}$. Similar to (4.11), the PDF is approximated by a multivariate Gaussian distribution with zero-mean, where for a small number of active antennas, the influence of the symbol vector on the signal received is small, i.e. $\mathrm{P}(\boldsymbol{y}|\boldsymbol{H}_i, \boldsymbol{s}) \approx \mathrm{P}(\boldsymbol{y}|\boldsymbol{H}_i)$. Note that with an increasing number of active antennas, the influence becomes more relevant and should be considered.

In analogy to the GAM for SM given in (4.11), the GAM for GMSM reads [LWL15]

$$\hat{\imath} = \underset{i \in \{1,\dots,N_\mathrm{p}\}}{\operatorname{argmax}} \ \frac{\exp\left(-\boldsymbol{y}^\mathsf{H}(\rho \boldsymbol{H}_i \boldsymbol{H}_i^\mathsf{H} + \boldsymbol{I}_{N_\mathrm{rx}})^{-1}\boldsymbol{y}\right)}{\det(\rho \boldsymbol{H}_i \boldsymbol{H}_i^\mathsf{H} + \boldsymbol{I}_{N_\mathrm{rx}})}. \tag{5.5}$$

In Section 4.1.2, we have proposed the modified GAM, where the computational complexity is reduced by applying the Sherman-Morrison formula (see, e.g. [GL12]). Applying this formula to (5.5) leads to the MGAM rule

$$\begin{aligned} \hat{\imath} = & \underset{i \in \{1,\dots,N_\mathrm{p}\}}{\operatorname{argmax}} \ \boldsymbol{y}^\mathsf{H} \boldsymbol{H}_i \boldsymbol{V}_i \boldsymbol{H}_i^\mathsf{H} \boldsymbol{y} - c_i, \\ & \text{with } \boldsymbol{V}_i = \rho \left(\boldsymbol{I}_{N_\mathrm{a}} + \rho \boldsymbol{H}_i^\mathsf{H} \boldsymbol{H}_i\right)^{-1} \\ & \text{and } c_i = \log\left(\det(\rho \boldsymbol{H}_i^\mathsf{H} \boldsymbol{H}_i + \boldsymbol{I}_{N_\mathrm{a}})\right). \end{aligned} \tag{5.6}$$

The computational complexity is reduced, since in contrast to (5.5), (5.6) requires the calculation of the determinant and inverse of a $N_\mathrm{a} \times N_\mathrm{a}$ matrix instead of a $N_\mathrm{rx} \times N_\mathrm{rx}$ matrix. For GMSM transmission scenarios, N_a is usually small compared to N_rx. Moreover, the term

$$\boldsymbol{V}_i \boldsymbol{H}_i^\mathsf{H} \boldsymbol{y} = (1/\rho \boldsymbol{I}_{N_\mathrm{a}} + \boldsymbol{H}_i^\mathsf{H} \boldsymbol{H}_i)^{-1} \boldsymbol{H}_i^\mathsf{H} \boldsymbol{y} \tag{5.7}$$

in (5.6) is equivalent to the block minimum mean-squared error (MMSE) term proposed in [XYD+14]. Furthermore, for high-SNR values we have

$$(1/\rho \boldsymbol{I}_{N_\mathrm{a}} + \boldsymbol{H}_i^\mathsf{H} \boldsymbol{H}_i)^{-1} \boldsymbol{H}_i^\mathsf{H} \boldsymbol{y} \approx (\boldsymbol{H}_i^\mathsf{H} \boldsymbol{H}_i)^{-1} \boldsymbol{H}_i^\mathsf{H} \boldsymbol{y}, \tag{5.8}$$

which is equivalent to the decorrelator (linear zero-forcing equalizer) described in [WJS12a]. Hence, the GAM is similar to both previously published methods except for the term c_i.

5.1.2 Detection of the Signal Points

First, we consider the same two-stage detection approach as for SM. In the first stage, we use the MGAM for an estimation of the antenna pattern. In the second stage, we calculate

$$\hat{\boldsymbol{s}} = \underset{\boldsymbol{s} \in \mathcal{X}}{\operatorname{argmin}} \, ||\boldsymbol{y} - \sqrt{\rho} \boldsymbol{H}_{\hat{i}} \boldsymbol{s}||^2 \tag{5.9}$$

for ML detection of the symbol. This procedure reduces the search complexity of the detection to order $\mathcal{O}\left(N_\mathrm{p} + M_2^{N_\mathrm{a}}\right)$. It can be seen, that the order still increases exponantially with the number of active transmit antennas. Hence, this fact motivates suboptimal detection techniques for the symbol detection. Note that the following approach can equally be applied to SM with $N_\mathrm{a} = 1$. In the case of SM, this becomes more relevant when the transmission scheme employ a signal set with high cardinality.

In the following, we assume again that the MGAM is applied in the first detection stage. In order to reduce the complexity of the symbol detection, we perform preprocessing using the pseudoinverse of the MIMO channel matrix. For further reduction, the matrices $\boldsymbol{V}_i$ calculated for detecting the antenna pattern are re-used, i.e. the MMSE linear equalization matrix

$$\boldsymbol{B}_i^\mathrm{MMSE} = \frac{1}{\sqrt{\rho}} \boldsymbol{V}_i \boldsymbol{H}_i^\mathsf{H} \tag{5.10}$$

Table 5.1 Estimation of the computational complexity for different detection methods

Detection method	MAC per received vector
ML	$N_{\text{rx}} N_{\text{p}} M_2^{N_{\text{a}}}$
MGAM $L = 2M_2^{N_{\text{a}}}$	$N_{\text{p}} N_{\text{a}} (N_{\text{rx}} + N_{\text{a}}) + 2M_2^{N_{\text{a}}} N_{\text{rx}}$
MGAM $L = N_{\text{a}}$	$N_{\text{p}} N_{\text{a}} (N_{\text{rx}} + N_{\text{a}}) + N_{\text{a}} N_{\text{rx}}$
MGAM $L = 2^{N_{\text{a}}+1}$	$N_{\text{p}} N_{\text{a}} (N_{\text{rx}} + N_{\text{a}}) + 2^{N_{\text{a}}+1} N_{\text{rx}}$

is utilized to equalize the channel matrix for the estimated antenna pattern. The estimate

$$\tilde{s} = \boldsymbol{B}_i^{\text{MMSE}} \boldsymbol{y} \tag{5.11}$$

enables independent detection for the transmitted symbols, where each element of $\tilde{s}$ is mapped to the nearest element in the constellation $\mathcal{X}$. This results in the estimate

$$\hat{s} = [\hat{s}_1, \ldots, \hat{s}_{N_{\text{a}}}]^{\mathsf{T}}. \tag{5.12}$$

Similar to the original GAM proposal in [LWL15], the detection performance can be improved by considering a list of antenna patterns. We consider only list-of-two detections, i.e.

$$\left\{\hat{i}^{(1)}, \hat{i}^{(2)}\right\} = \underset{i \in \{1,\ldots,N_{\text{p}}\}}{\text{argsort}} \ \tilde{\boldsymbol{y}}_i^{\mathsf{H}} \boldsymbol{V}_i \tilde{\boldsymbol{y}}_i - c_i, \tag{5.13}$$

where $\tilde{\boldsymbol{y}}_i = \boldsymbol{H}_i^{\mathsf{H}} \boldsymbol{y}$, and the argsort operation sorts the values in descending order and returns the indices of the best and second-best candidates, denoted as $\hat{i}^{(1)}$ and $\hat{i}^{(2)}$, respectively. Hence, we calculate $\boldsymbol{B}_{\hat{i}^{(1)}}^{\text{MMSE}}$ and $\boldsymbol{B}_{\hat{i}^{(2)}}^{\text{MMSE}}$ to equalize the channel matrices. The transmitted symbol vector is detected as

$$\left(\hat{i}, \hat{s}\right) = \underset{(i,s) \in \{(\hat{i}^{(1)}, \hat{s}^{(1)}), (\hat{i}^{(2)}, \hat{s}^{(2)})\}}{\text{argmin}} ||\boldsymbol{y} - \sqrt{\rho} \boldsymbol{H}_i \boldsymbol{s}||^2, \tag{5.14}$$

where $\hat{s}^{(1)}, \hat{s}^{(2)}$ are the detected symbol vectors corresponding to the two best antenna patterns.

Furthermore, we can generate more candidates by considering the best and second-best candidate for every symbol decision. In this case, we obtain a list with $L = 2^{N_{\text{a}}+1}$ candidates, cf. [RSF+19].

The overall search complexity of this approach is $\mathcal{O}\left(N_{\text{p}} + N_{\text{a}} M_2 + L\right)$, because N_{p} comparisons are required for detecting the best and second-best antenna patterns.

Similarly, the symbol detection requires N_a searches in constellations with M_2 elements. Typically, with GMSM, the number of active antennas is small. Hence, even the final detection step with a list size $L = 2^{N_a+1}$ has a minor contribution to the total search complexity.

5.1.3 Computational Complexity Analysis

In this section, we analyze the computational complexity of the proposed suboptimal decoding method. We determine the computational complexity in terms of MAC operations. Moreover, we compare numerical results for the proposed method with results for ML detection and the MGAM with list-of-two detections.

For all methods, the leading terms for the computational complexity are shown in Table 5.1. These terms consider only the detection complexity per received vector, i.e. the computational complexity to estimate the channel is neglected, since it only needs to be computed once for each new channel instance. However, the computations per channel instance are discussed in Section 5.2.3.

For ML detection, (5.3) has to be evaluated for all $N_p M_2^{N_a}$ possible transmit vectors, where each matrix-vector product $\boldsymbol{H}_i \boldsymbol{s}$ requires $N_a N_{rx}$ MAC operations. However, these matrix-vector products can be calculated once for each new channel instance. Hence, for each received vector, only the difference and vector norm have to be calculated, which requires only N_{rx} MAC operations per transmit vector. Consequently, the leading term for ML detection in Table 5.1 is $N_{rx} N_p M_2^{N_a}$.

The MGAM method is a multi-stage detection procedure. First, the antenna pattern is estimated according to (5.13). This requires two matrix-vector products with $N_a N_{rx}$ and N_a^2 MAC operations per antenna pattern, respectively. Hence, the antenna pattern detection requires $N_p N_a (N_{rx} + N_a)$ operations. The second stage is the signal-vector detection, which requires N_{rx} MAC operations per transmit vector in the list. The total number of operations depends on the list size L. The formulas for the different list sizes are provided in Table 5.1. Note that argsort operations, e.g. in (5.13) are logical comparisons with the complexity $\mathcal{O}\left(N_p\right)$ and they require no additional MAC operations.

Figures 5.1 and 5.2 show numerical results for the computation complexity with $N_{tx} = 16$ transmit and $N_{rx} = 8$ receive antennas for $N_a = 2$ and $N_a = 3$, respectively. As can be seen from Table 5.1 and the figures, the computational complexity with the proposed suboptimal decoding methods (list size $L = N_a$ or $L = 2^{N_a+1}$) is independent of M_2, whereas the complexity of ML and the original MGAM method (list size $L = 2M_2^{N_a}$) increases with order $\mathcal{O}\left(M_2^{N_a}\right)$.

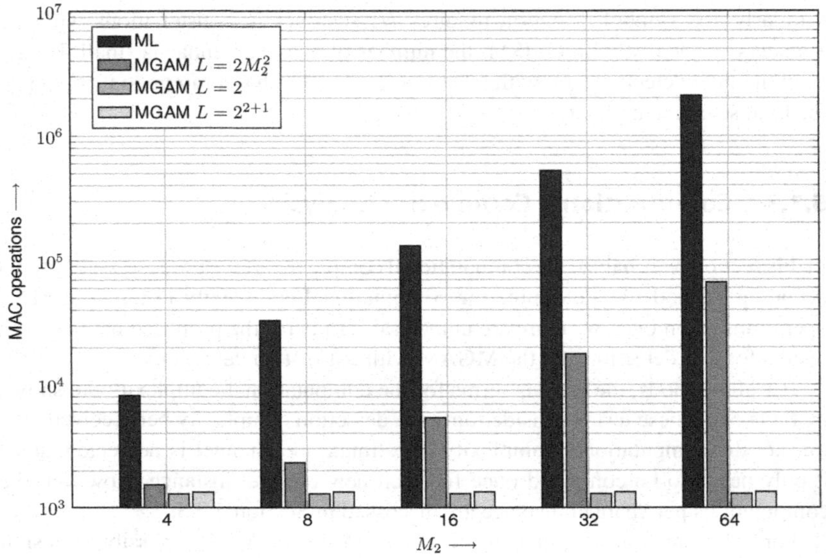

Figure 5.1 Computation complexity in MAC operations for different decoding methods depending on the alphabet size M_2 with $N_p = 2^{\lfloor \log_2 \binom{N_{tx}}{N_a} \rfloor} = 64$ antenna patterns

5.1.4 Simulation Results

Next, we show results of Monte-Carlo simulations. For the sake of brevity, we consider only a single transmission scheme with $N_{tx} = 12$ and $N_{rx} = 8$. In the simulations, 10^7 symbol vectors were randomly generated per SNR value and transmitted over the channel. Detection was performed using the noisy signal samples received, assuming perfect CSI at the receiver.

Figures 5.3 and 5.4 present results for two and three active transmit antennas, respectively. With $N_a = 2$, we use 8 QAM with $M_2 = 8$ elements. For the case of $N_a = 3$, we use a square QAM constellation with $M_2 = 4$ elements. In both cases, we have $N_p = 16$ antenna patterns, where we use only 16 antenna patterns to reduce the complexity with ML detection and obtain the same spectral efficiency of 10 bpcu for two and three active antennas. As can be seen in Figure 5.3, ML detection has the best performance, the MGAM method with suboptimal detection and the proposed detection approach achieve practically the same performance. ML

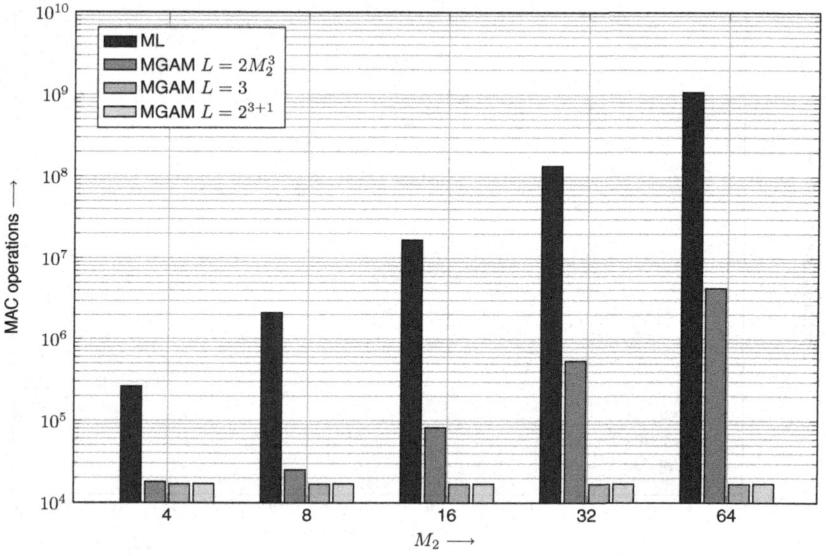

Figure 5.2 Computation complexity in MAC operations for different decoding methods depending on the alphabet size M_2 with $N_p = 2^{\lfloor \log_2 \binom{N_{tx}}{N_a} \rfloor} = 512$ antenna patterns

has to consider all 2^{10} possible transmit vectors. The MGAM approach requires 16 comparisons to detect the antenna pattern, whereas $2M_2^2 = 128$ comparisons for the signal vectors are required because the best and second-best antenna patterns are considered. With the proposed method, only $2^{2+1} = 8$ signal points are considered for the symbol detection.

Figure 5.4 presents results for $N_a = 3$ active transmit antennas. The spectral efficiency is also 10 bpcu. In this case, the MGAM detection also results in a small loss compared with ML detection. Similarly, the proposed method exhibits the same performance loss, which is caused by erroneous detection of the antenna pattern. On the other hand, the proposed MGAM with reduced L achieves the same performance as the MGAM method with the maximum L size. Hence, the reduced complexity for the symbol detection does practically not induce additional detection errors compared with the MGAM method.

It is also interesting to compare the results of Figures 5.3 and 5.4, where the two GSM schemes have the same spectral efficiency. GSM with three active antennas and 4 QAM outperforms the GSM scheme with two active antennas and 8 QAM.

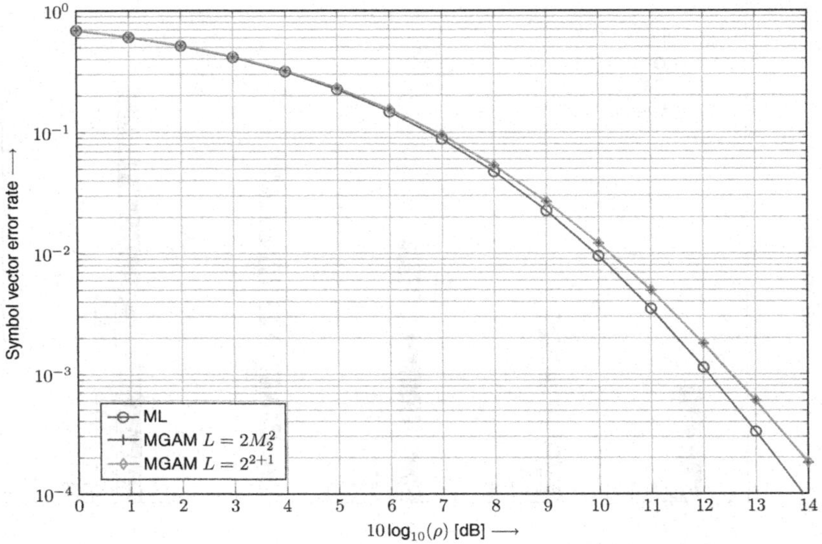

Figure 5.3 Simulation results for GMSM using two 8 QAM and transission over two active transmit antennas. The suboptimal detection approaches applying the MGAM to estimate the best and second-best antenna pattern. For the second stage, a different number of candidates are considered acc. to L

However, increasing the number of active antennas may have issues that are not considered in the channel model, e.g. inter-channel interference introduced by coupling multiple symbols in time and space. Moreover, we can observe that the performance loss compared with ML detection increases, when more active transmit antennas are used. Simulation results using more than three active transmit antennas have shown the same behaviour with an increasing error probability for the estimation of the antenna pattern resulting in an overall worser error performance. Despite other reasons, one is that the GAM and the MGAM approximate the received vector with zero-means multivariate Gaussian distributions, where the sent symbol(s) influence the variances and hence the distributions overlap more. If we consider one or two active transmit antennas, the influence of the sent symbol on the received signal is usually small compared to the case of $N_\mathrm{a} \geq 3$. Hence, with more active transmit antennas, the approximation error increases.

5.2 Hurwitz-Integer Signal Constellations for GMSM

The following part of this chapter focuses on four-dimensional signal constellations, with elements from the Hurwitz integers, and their application in GMSM transmission schemes. Note, that the construction of the signal sets differ from the one described in Section 3.2.2. There, finite sets are created by the definition and application of a modulo function. Here, we follow a different approach. We define a limitation of the highest norm for the elements and hence obtain a four-dimensonal sphere. Even though the construction is quite different, we can use the set partitioning method from Section 3.2.4.

5.2.1 Construction of Finite Sets

In order to construct signal constellations as subsets of the Hurwitz integers, we use $s_1 = a_1 + a_2\mathrm{i}$ for the first active antenna and $s_2 = a_3 + a_4\mathrm{i}$ for the second one. This construction results in two projections $\mathcal{H} \rightarrow \mathbb{C}$, where the first projection

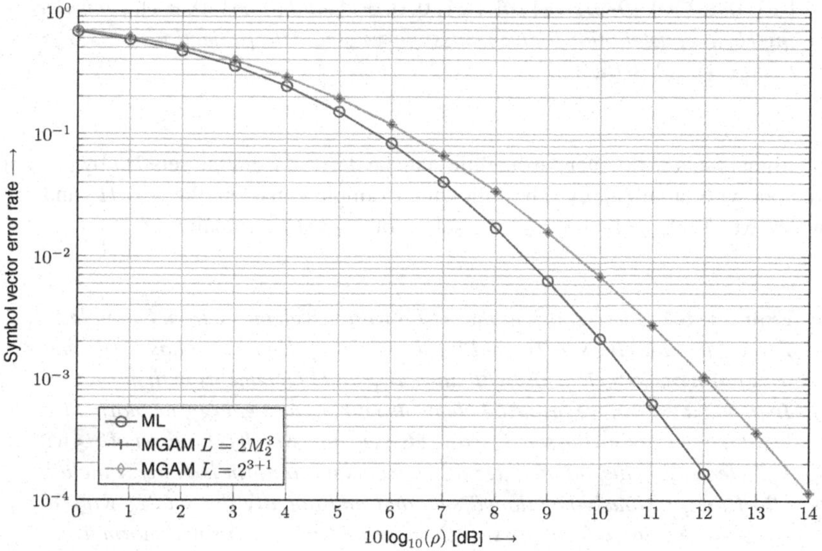

Figure 5.4 Simulation results for GSM with three active antennas and 4 QAM

defines the signal set $\mathcal{T}_1$ for the first antenna and the second one the set $\mathcal{T}_2$ for the second antenna. For GMSM transmission, every active antenna has to send a non-zero signal; otherwise, the antenna pattern would not be unique. The zeros indicate inactive antennas. Hence, we use only Hurwitz integers from the set

$$\tilde{\mathcal{H}} = \mathscr{H} \backslash \{h \in \mathscr{H} \mid a_1 = a_2 = 0 \vee a_3 = a_4 = 0\} \tag{5.15}$$

for GMSM transmission. Finite constellations are given by all points where the norm is limited to $d \in \mathbb{N}$ with

$$\tilde{\mathcal{H}}_d = \{h \in \tilde{\mathcal{H}} \mid N(h) \leq d\}. \tag{5.16}$$

Some of these sets have a cardinality, which is a power of two, e.g. $|\tilde{\mathcal{H}}_1| = 2^4$, $|\tilde{\mathcal{H}}_2| = 2^5$ and $|\tilde{\mathcal{H}}_3| = 2^7$.

Example 5.1 *The set $\tilde{\mathcal{H}}_1$ is the set of 16 Hurwitz integers of the form $[\pm\frac{1}{2}, \pm\frac{1}{2}, \pm\frac{1}{2}, \pm\frac{1}{2}]$. The set $\tilde{\mathcal{H}}_2$ contains 32 elements including $\tilde{\mathcal{H}}_1$ and all 16 integers of the form* $[\pm 1, 0, \pm 1, 0]$, $[0, \pm 1, \pm 1, 0]$, $[\pm 1, 0, 0, \pm 1]$, *and* $[0, \pm 1, 0, \pm 1]$. *Considering $\tilde{\mathcal{H}}_2$ as constellation for transmission, we obtain the projections $\mathcal{T}_1 = \mathcal{T}_2 =$* $\{\pm\frac{1}{2} \pm \frac{1}{2}\mathrm{i}, \pm 1 + 0\mathrm{i}, 0 \pm 1\mathrm{i}\}$.

Signals sets with other cardinalities can be derived from the sets $\tilde{\mathcal{H}}_d$ by pruning vectors with large energy. The following example considers the sets $\tilde{\mathcal{H}}_3^6$ and $\tilde{\mathcal{H}}_5^8$, where we use the notation $\tilde{\mathcal{H}}_d^b$ for a subset of $\tilde{\mathcal{H}}_d$ with cardinality 2^b.

Example 5.2 *First, consider the set $\tilde{\mathcal{H}}_3$ with* 128 *elements from Example 5.1. This set includes the sets $\tilde{\mathcal{H}}_1$ and $\tilde{\mathcal{H}}_2$, respectively. Hence, $\tilde{\mathcal{H}}_3$ has 16 elements with norm one, 16 elements with norm two and 96 elements with norm three. To form the set $\tilde{\mathcal{H}}_3^6$ we prune 64 elements with highest energy, i.e. norm three. Similarly, we obtain the set $\tilde{\mathcal{H}}_5^8$ with* 256 *elements, which is a subset of $\tilde{\mathcal{H}}_5$ with* 272 *elements. This set includes all elements up to norm five, i.e. it includes all of the previously-mentioned sets and additionally the set $\tilde{\mathcal{H}}_4$ with* 144 *elements. Again, to form the set $\tilde{\mathcal{H}}_5^8$ we prune* 16 *elements with norm five.*

5.2.2 Suboptimal Symbol Detection with Hurwitz-Based Signal Sets

Next, we consider the particular partitioning of the Hurwitz integers, which is used for the symbol detection approach. This technique exploits the dependencies between the elements of the projections $\mathcal{T}_1$ and $\mathcal{T}_2$, similar to the set partitioning technique in Section 3.2.4. For instance, a Hurwitz integer has all elements either from the set of half-integers or the set of full-integers. Hence, if we select an integer from $\mathcal{T}_1$, this selections restricts the possible elements from $\mathcal{T}_2$ to the integer elements. Such dependencies can be used to simplify the detection procedure. First detecting the elements from one antenna reduces the search space for detecting the signal points of the second antenna. The subset of $\mathcal{T}_1$, which is determined by fixing an element $s \in \mathcal{T}_2$, is denoted as $\mathcal{T}_1(s)$. Similarly, $\mathcal{T}_2(s) \subset \mathcal{T}_2$ is determined by an element $s \in \mathcal{T}_1$, which we illustrate with the following example.

Figure 5.5 depicts a more complicated example where the signal set $\mathcal{T}_1$ is the projection of $\tilde{\mathcal{H}}_3^6$. Choosing one symbol $s \in \mathcal{T}_1$ determines one of five possible subsets of $\mathcal{T}_2$. Any symbol with same color/marker leads to the same subset. Note that the partitioning of $\mathcal{T}_2$ due to fixing a point in $\mathcal{T}_1$ results in the same subsets.

Example 5.3 *Consider the set $\tilde{\mathcal{H}}_2$ from Example 5.1, which contains all half- and full-integers with norm one or two. Any point $s \in \{\pm\frac{1}{2} \pm \frac{1}{2}\mathrm{i}\}$ results in the subset of half-integers, e.g. $\mathcal{T}_1\left(\frac{1}{2} - \frac{1}{2}\mathrm{i}\right) = \mathcal{T}_2\left(\frac{1}{2} - \frac{1}{2}\mathrm{i}\right) = \{\pm\frac{1}{2} \pm \frac{1}{2}\mathrm{i}\}$. Similarly, any point from the set $\{\pm 1 + 0\mathrm{i}, 0 \pm 1\mathrm{i}\}$ has the consequence that the points in the subsets are also full integers, e.g., $\mathcal{T}_1(-1) = \mathcal{T}_2(-1) = \{\pm 1 + 0\mathrm{i}, 0 \pm 1\mathrm{i}\}$.*

In Section 5.1.2, we have discussed a suboptimal detection approach for GMSM, which is able to estimate the closest points to a received one when an arbitrary multidimensional signal set is applied. Usually the sets are based on conventional two-dimensional signal constellations, i.e. the set $\mathcal{X} = \mathcal{T}^{N_\mathrm{a}}$ is the N_a-fold Cartesian product of the complex-valued signal constellation $\mathcal{T}$.

Multidimensional signal constellations based on the densest known lattices may result in higher CFM values and thus lead to a better error performance on the AWGN channel, when compared with the aforementioned conventional signal sets. On the other hand, in Figure 5.5, we can observe that the first search in $\mathcal{T}_1$ is higher, but reduces in the second projection $\mathcal{T}_2$.

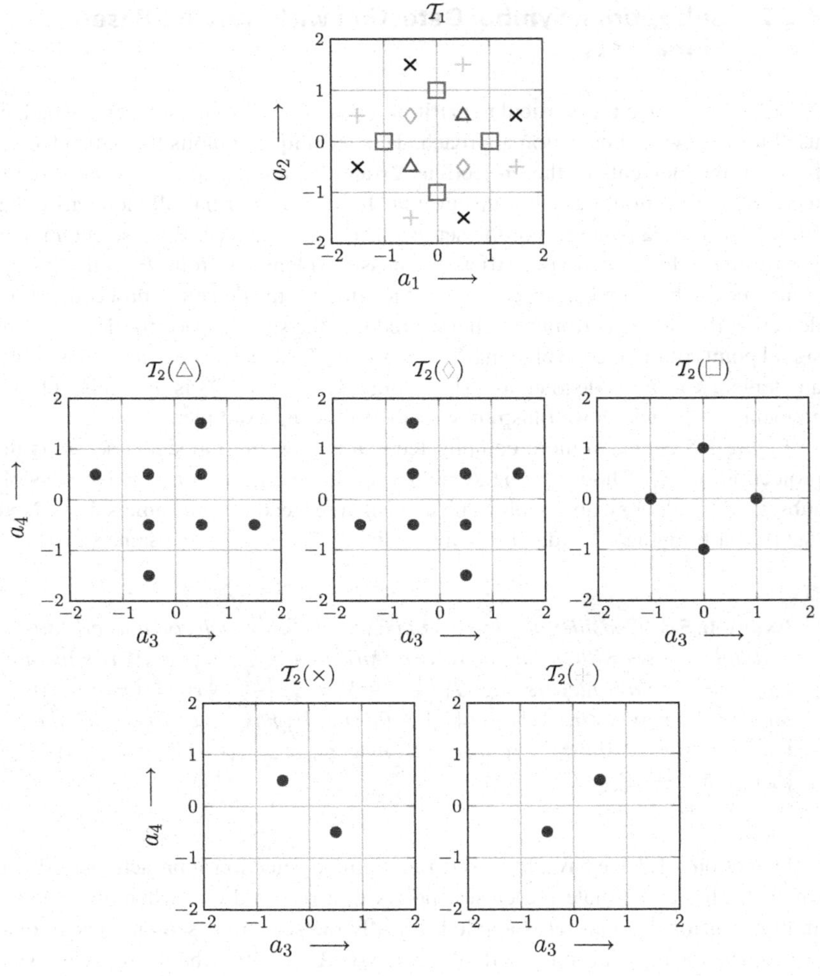

Figure 5.5 The uppermost plot depicts the projection $\mathcal{T}_1$ from $\tilde{\mathcal{H}}_3^6$ and is used for the first symbol decision. All possible subsets are depicted in plots in the second and third plot row. Choosing one point in $\mathcal{T}_1$ leads to a specific subset in $\mathcal{T}_2$. Any point with the same color/marker in $\mathcal{T}_1$ shares the same subset in $\mathcal{T}_2$

In order to reduce the search space for the symbol detection, we exploit the set partitioning of the Hurwitz integers. Similar to the approach presented in Section 5.1.2, we can utilize a linear equalizer to obtain an estimate $\tilde{\boldsymbol{s}} = \boldsymbol{B}_i^{\mathrm{MMSE}} \boldsymbol{y}$ of the submitted symbol vector $\boldsymbol{s} = [s_1, s_2]^{\mathsf{T}}$ with $s_1 \in \mathcal{T}_1$ and $s_2 \in \mathcal{T}_2$. However, for a fixed value of s_1, we have $s_2 \in \mathcal{T}_2(s_1)$. Similarly, for a fixed value of s_2, we have $s_1 \in \mathcal{T}_1(s_2)$. We use this fact to determine two candidate vectors $\hat{\boldsymbol{s}}' = \left[\hat{s}_1', \hat{s}_2'\right]^{\mathsf{T}}$ and $\hat{\boldsymbol{s}}'' = \left[\hat{s}_1'', \hat{s}_2''\right]^{\mathsf{T}}$. The symbol $\hat{s}_1'$ is determined as the symbol from $\mathcal{T}_1$, which minimizes the Euclidean distance to the MMSE estimate $\tilde{s}_1$. Next, $\hat{s}_2'$ is determined as the symbol from $\mathcal{T}_2(\hat{s}_1')$ based on the MMSE estimate $\tilde{s}_2$. Similarly, $\hat{\boldsymbol{s}}''$ is obtained by first detecting $\hat{s}_2'' \in \mathcal{T}_2$ and subsequently $\hat{s}_1'' \in \mathcal{T}_1(\hat{s}_2'')$.

For large signal sets, the size of the projections is much smaller than the cardinality of the Hurwitz constellation. Consequently, the total number of comparisons can be reduced. We present numerical results for the complexity in the next section.

Further candidates for the transmitted symbol vector can be obtained by considering more candidates for the antenna pattern. As before, we consider two candidates $\hat{\imath}^{(1)}, \hat{\imath}^{(2)}$ for the antenna index. Finally, the symbol vector is determined as

$$(\hat{\imath}, \hat{\boldsymbol{s}}) = \underset{\substack{(i,\boldsymbol{s}) \in \left\{ (\hat{\imath}^{(1)}, \hat{\boldsymbol{s}}^{(1),\prime}), (\hat{\imath}^{(2)}, \hat{\boldsymbol{s}}^{(2),\prime}), \right. \\ \left. (\hat{\imath}^{(1)}, \hat{\boldsymbol{s}}^{(1),\prime\prime}), (\hat{\imath}^{(2)}, \hat{\boldsymbol{s}}^{(2),\prime\prime}) \right\}}}{\operatorname{argmin}} ||\boldsymbol{y} - \sqrt{\rho} \boldsymbol{H}_i \boldsymbol{s}||^2. \tag{5.17}$$

5.2.3 Computational Complexity Analysis

In this subsection, we analyze the computational complexity of the proposed suboptimal symbol detection approach for Hurwitz-based signal constellations from Section 5.2.2. We compare numerical results for the proposed method with results for the generalized suboptimal detection method given in Section 5.1.3.

We re-use the results given in Table 5.1 for ML detection, which requires $N_{\mathrm{rx}} N_{\mathrm{p}} M_2^{N_{\mathrm{a}}}$ MAC operations per received vector.

In the following, we consider the number of MAC operations per channel instance for the suboptimal detection according to Section 5.1.3. The detection of the antenna pattern in (5.6) results in $N_{\mathrm{p}}(N_{\mathrm{rx}}^2 + N_{\mathrm{rx}})$ operations, i.e. each of the N_{p} patterns require N_{rx}^2 MAC operations for a matrix-vector product and N_{rx} for the scalar product. The second stage in the MGAM detection is the symbol-vector detection, which requires N_{rx} MAC operations per transmit vector for each candidate. The total number of operations depends on the total number of candidates. The original GAM [LWL15] and the MGAM from Section 5.1.3 consider all $M_2^{N_{\mathrm{a}}}$ possible symbol vectors $s \in \mathcal{X}$ in (5.9) for the best and second-best antenna pattern estimates, hence we have $L = 2M_2^{N_a}$ candidates. We denote this approach as MGAM-ML detection, because the symbol vector detection applies the ML rule. This MGAM-ML approach results in $2M_2^{N_a} N_{rx}$ MAC operations.

The proposed suboptimal approach, denoted as MGAM-SD, requires additional computations for channel equalization once per channel instance. These computations are not considered. For each vector received, the linear MMSE estimate $\tilde{s}$ needs to be computed. This matrix-vector product requires $N_a N_{rx}$ MAC operations. In addition, for all N_a symbols the search for the nearest and second-nearest symbols within the constellation has to be performed, which requires $N_a M_2$ MAC operations. The evaluation of (5.14) with the extension to consider the best and second-best candidate for every symbol decision leads to $L = 2^{N_a+1}$ candidates, which requires $2^{N_a+1} N_{rx}$ MAC operations. The overall computational complexity per received vector for MGAM-SD has a total number of $2N_a(N_{rx}+M_2)+2^{N_a+1}N_{rx}$ MAC operations per received vector when the best and second-best antenna patterns are considered.

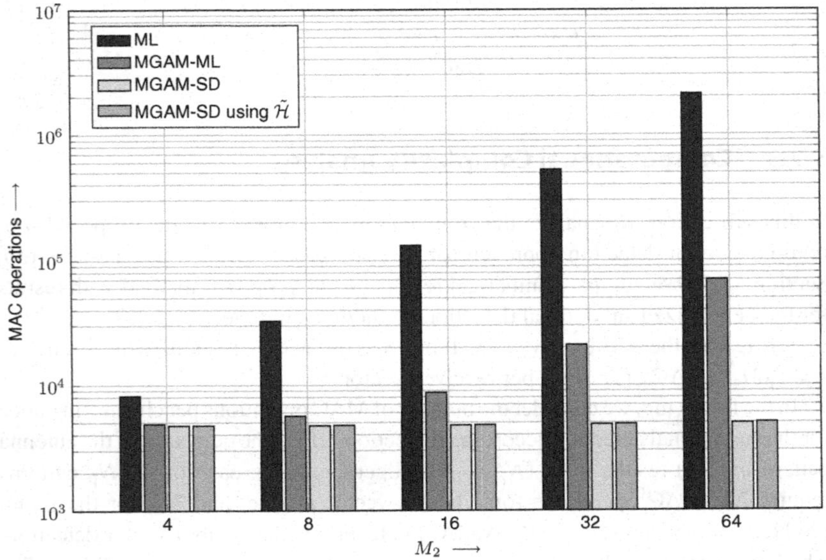

Figure 5.6 Computational complexity in MAC operations for symbol detection depending on the alphabet size M_2 with $N_p = 64$ antenna patterns and $N_a = 2$ active antennas

Figure 5.6 shows numerical results for the total computational complexity per received vector for $N_{\text{tx}} = 16$ transmit and $N_{\text{rx}} = 8$ receive antennas. The signal constellations for ML, MGAM-ML, and MGAM-SD are based on the Cartesian products of two independent complex-valued M_2-ary constellations. For MGAM-SD using $\tilde{\mathcal{H}}$, the corresponding Hurwitz constellation sets $\tilde{\mathcal{H}}_d^{2\log_2(M_2)}$, with their related projections according to Section 5.2.2, are considered. Hence, for a certain M_2, the spectral efficiency for the M_2-ary constellation and the $\tilde{\mathcal{H}}_d^{2\log_2(M_2)}$ Hurwitz constellation is equal. For the Hurwitz constellations, the sizes of the projections vary. The numerical results take the corresponding set sizes into account. The alphabet size M_2 has only a minor impact on the computational complexity with the proposed suboptimal detection method MGAM-SD (list size $L = 2^{N_a+1}$), whereas the complexity of ML and the original MGAM-ML method from [FRS18] ($L = 2M_2^{N_a}$) increases with order $\mathcal{O}\left(M_2^{N_a}\right)$. The results in Figure 5.6 demonstrate that the MGAM-SD detection of the Hurwitz constellations has practically the same complexity as the detection of independent symbols.

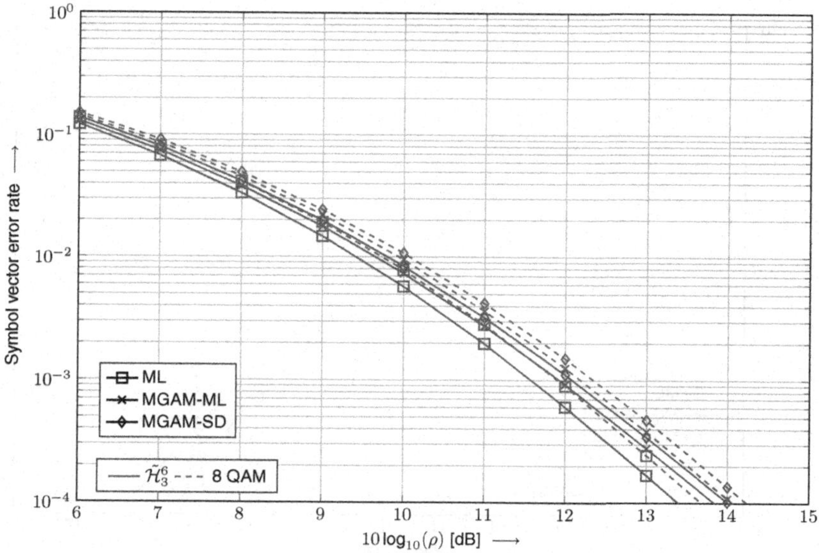

Figure 5.7 Simulation results for GMSM with two active antennas. The spectral efficiency is 10 bpcu

5.2.4 Simulation Results

Finally, we present results obtained from Monte-Carlo simulations for transmission with GMSM from Section 2.2.3. For the sake of brevity, we consider only a single transmission scheme with $N_{\mathrm{tx}} = 8$ and $N_{\mathrm{rx}} = 8$. With $N_{\mathrm{a}} = 2$, this setup results in $\binom{8}{2} = 28$ possible antenna patterns. With a binary mapping $\lfloor \log_2(28) \rfloor = 4$ bits can be transmitted using $N_{\mathrm{p}} = 16$ antenna patterns. These 16 patterns were randomly selected and the same set was used for all simulations. Furthermore, 10^7 symbol vectors were randomly generated per SNR value and transmitted over the channel. Detection was performed using the received noisy signal samples, assuming perfect CSI at the receiver.

Figure 5.7 present results for two active transmit antennas with a spectral efficiency of 10 bpcu. For the Cartesian product of two independent alphabets, we use 8 QAM. As can be seen in Figure 5.7, the Hurwitz constellation in combination with ML detection has the best performance. The suboptimal detection methods lead to a small performance loss. With the proposed method (MGAM-SD), the

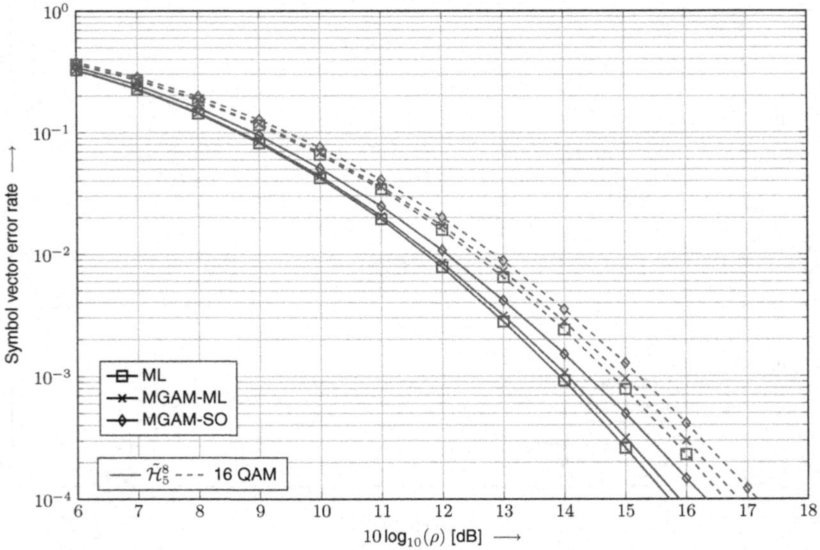

Figure 5.8 Simulation results for GMSM with two active antennas. The spectral efficiency is 12 bpcu

Hurwitz constellation and the Cartesian product of 8 QAM have nearly the same performance.

For larger signal sets, i.e. higher spectral efficiencies, the gain of the Hurwitz integers becomes more significant. This is demonstrated in Figure 5.8, which presents results for GMSM with 12 bpcu. The MGAM-ML achieves practically ML performance, whereas the proposed MGAM-SD method leads to a small loss with both constellations. However, the Hurwitz constellation with the proposed MGAM-SD detection method outperforms the Cartesian product of two independent conventional 16 QAM constellations with ML detection. In this case, the multidimensional constellation leads to a performance gain that can be achieved with a smaller computational complexity compared with conventional two-stage GMSM detection.

5.3 OMEC Code Based GMSM Transmission

In Section 4.2 [RSF20], a space-time coding scheme is proposed that uses only one active transmit antenna per time instance, but forms a code over time employing OMEC codes and a set of transmission patterns. In contrast to Section 4.2 [RSF20], we consider the OMEC codes $\mathcal{C}$ as multidimensional signal constellations, i.e. $\mathcal{X} = \mathcal{C}$, for GMSM transmission and apply no coding over time. Furthermore, we utilize a suboptimal decoding strategy for these codes. This suboptimal decoding algorithm was introduced in [FGS13] for OMEC codes over Gaussian integers for the AWGN channel. In Section 3.1.2, we have generalized this approach to Eisenstein integers. In the following, we revise this detection approach, such that it results in a low-complexity detection algorithm for the proposed GMSM transmission scheme. At the end, simulation results show that the proposed method can outperform conventional GMSM with respect to decoding performance, detection complexity, and spectral efficiency.

5.3.1 Decoding and Detection

The code length is choosen according to the number of active antennas, i.e. $\mathrm{N}_c = N_\mathrm{a}$ with $N_\mathrm{a} \geq 2$. (3.16) and (3.17) imply a field size $p \geq 4N_\mathrm{a} + 1$ for Gaussian integers and $p \geq 6N_\mathrm{a} + 1$ for Eisenstein integers.

In order to use OMEC codes for GMSM transmission, we have to prune the code. The finite Gaussian-integer and Eisenstein-integer fields contain the zero element. However, with GMSM, the zero element should be omitted to ensure that the transmitted antenna pattern can be uniquely determined at the receiver. Hence, we use only a subcode $\mathcal{C}' \subset \mathcal{C}$, where we omit all codewords from the OMEC code $\mathcal{C}$ with elements that are zero. For instance, for the code from Example 3.2, this pruning reduces the number of codewords from $p^2 = 169$ to 132. In order to obtain a constellation where the number of elements is a power of two, we can additionally omit four codewords.

The pruned code $\mathcal{C}'$ is not a linear code, but it is a subcode of a linear code $\mathcal{C}$. We will use this fact for the detection. We apply the list decoding for the linear code $\mathcal{C}$ and prune the list of candidate codewords $\mathcal{L}$. Any candidate that contains zero elements is expunged from the list, which leads to the pruned list $\mathcal{L}' \subset \mathcal{C}'$.

In order to reduce the computational complexity of the GMSM detection, we use the pseudoinverse of the MIMO channel matrix to equalize the channel similar as described in Section 5.1.2 and [RSF+19]. Specifically, we apply the MMSE linear-equalization matrix

$$\boldsymbol{B}_i^{\text{MMSE}} = \sqrt{\rho}\left(\boldsymbol{I}_{N_{\text{a}}} + \rho \boldsymbol{H}_i^{\mathsf{H}} \boldsymbol{H}_i\right)^{-1} \boldsymbol{H}_i^{\mathsf{H}} \tag{5.18}$$

to equalize the channel matrix for every antenna pattern. The estimate

$$\boldsymbol{y}_i = \boldsymbol{B}_i^{\text{MMSE}} \boldsymbol{y} \tag{5.19}$$

will be used as the received vector for the list decoding of the OMEC codes. Let

$$\tilde{\boldsymbol{y}}_i = \lfloor \boldsymbol{y}_i \rceil, \tilde{y}_{i,j} \in \mathcal{G}_p \tag{5.20}$$

or

$$\tilde{\boldsymbol{y}}_i = [\boldsymbol{y}_i], \tilde{y}_{i,j} \in \mathcal{E}_p \tag{5.21}$$

be the hard-input vector at the receiver, i.e. the elements of $\tilde{\boldsymbol{y}}_i$ are mapped to the closest symbol within the Gaussian-integer or Eisenstein-integer constellation with respect to the squared Euclidean distance given in (3.39). We first determine the hard-input vector and calculate the syndrome

$$s_i = \boldsymbol{P}\tilde{\boldsymbol{y}}_i^{\mathsf{T}}. \tag{5.22}$$

Based on this syndrome, we calculate the pruned list $\mathcal{L}_i'$ and determine the best codeword for the ith antenna pattern

$$\hat{\boldsymbol{x}}_i = \underset{\boldsymbol{x} \in \mathcal{L}_i'}{\operatorname{argmin}} ||\boldsymbol{y}_i - \boldsymbol{x}||^2. \tag{5.23}$$

Finally, we estimate the best codeword and the antenna pattern as

$$(\hat{i}, \hat{\boldsymbol{x}}) = \underset{i \in \{1,\ldots,N_\mathrm{p}\}}{\operatorname{argmin}} ||\boldsymbol{y} - \boldsymbol{H}_i \hat{\boldsymbol{x}}_i||^2. \tag{5.24}$$

Note that this procedure reduces the number of vector comparisons to order $\mathcal{O}(N_\mathrm{p}L)$, where L is the size of the list $\mathcal{L}_i'$. It is shown in [FGS13] that the list size L grows linearly with the code length N_c and is independent of the field size p. Hence, with $N_\mathrm{c} = N_\mathrm{a}$, the number of vector comparisons is of order $\mathcal{O}(N_\mathrm{p}N_\mathrm{a})$ and independent of the size of the signal constellation. For the sake of brevity, we assume that the decoding procedure uses all N_p possible antenna patterns, but the computational complexity could be reduced by first applying a detection procedure for the antenna pattern [WJS12a, XYD+14, LWL15], e.g. MGAM, which could enable a reduced search using only likely antenna patterns.

5.3.2 Simulation Results

Figure 5.9 presents results for GMSM transmission with $N_\mathrm{tx} = 12$ transmit antennas, $N_\mathrm{rx} = 8$ receive antennas, and $N_\mathrm{a} = 3$ active antennas, where we can use up to $N_\mathrm{p} = 128$ antenna patterns. The simulations consider 10^4 randomly-chosen Rayleigh fading channel instances with 10^6 transmitted symbols per SNR value. We compare the performance of the OMEC codes with that of GMSM with conventional QAM and PSK constellations. For these conventional GMSM schemes, we choose transmission schemes with 14 bpcu and ML detection. All results for OMEC codes are obtained with the proposed suboptimal detection scheme. For the codes over Eisenstein integers, we have $K = 256$ codewords and the spectral efficiencies 14 or 15 bpcu with 64 or 128 antenna patterns, respectively. Note that the GMSM scheme with $\mathcal{G}_{13}$, $K = 128$, and $N_\mathrm{p} = 128$ has 14 bpcu and practically the same performance as the transmission scheme with $\mathcal{E}_{19}$ and 14 bpcu. The transmission scheme with $\mathcal{E}_{19}$ and 15 bpcu still outperforms the conventional GMSM schemes with 14 bpcu based on 8 QAM or 8 PSK with ML detection. Note that the proposed Eisenstein transmission scheme with 15 bpcu requires only $LN_\mathrm{p} = 1280$ vector comparisons per received vector, whereas ML detection for the conventional scheme requires $2^{14} = 16384$ vector comparisons.

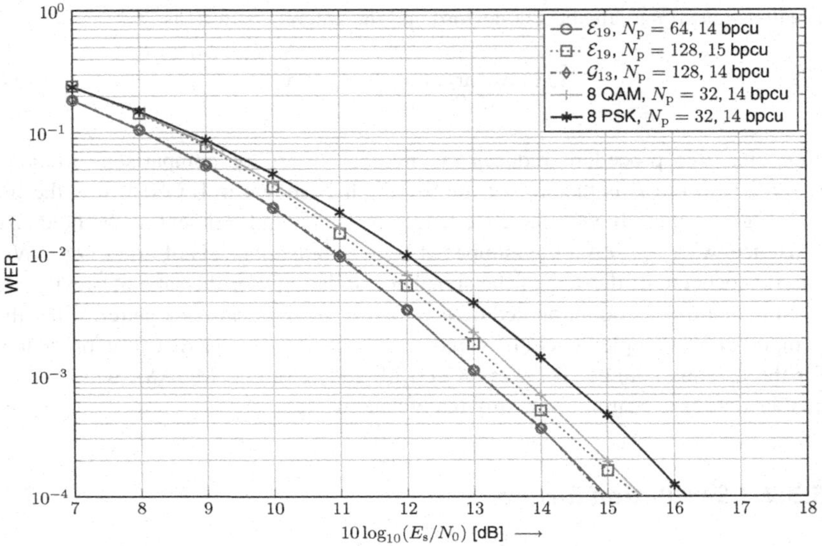

Figure 5.9 Word error rate (WER) versus SNR in dB for Eisenstein ($\mathcal{E}_{19}$), Gaussian ($\mathcal{G}_{13}$), and conventional (8 QAM and 8 PSK) constellations. GMSM transmission with $N_{tx} = 12$ transmit antennas, $N_{rx} = 8$ receive antennas, and $N_a = 3$ active antennas. ML detection is performed on the received noisy signal samples from the conventional signal constellations and the proposed suboptimal detection is applied for the Eisenstein and Gaussian constellation(s)

5.4 Discussion

In the first part of this chapter, we have discussed a suboptimal detection approach for GMSM transmission schemes, where the suboptimal detection procedure is separated into two-stages: The first stage to estimate the antenna pattern and the second stage to estimate the transmitted symbols. For the first stage, we have revised the Gaussian approximation method [LWL15] for the detection of the antenna pattern. In Section 4.1.2, we have shown the modification of the Gaussian approximation method for one active transmit antenna. In Section 5.1.1, the approach was generalized for the case of more than one active transmit antenna. Subopotimal detection for the antenna pattern reduces the computational complexity from order $\mathcal{O}\left(N_p M_2^{N_a}\right)$ for overall-ML detection to order $\mathcal{O}\left(N_p + M_2^{N_a}\right)$, with $\mathcal{O}\left(N_p\right)$ for the first and

$\mathcal{O}\left(M_2^{N_a}\right)$ for the second stage. Note, that the complexity order of the second stage still rises exponentially with the number of active transmit antennas. We have presented a suboptimal detection approach for the second detection stage, where indedependent symbol detection is enabled. It is based on linear equalization of the channel matrix. With this approach, the complexity order is significantly reduced and just rises linear with the number of active transmit antennas. Moreover, it can be applied in case of one active antenna, which may become more important when high-modulation orders are present.

The presented simulation results indicate, that the same performance is achieved as with ML detection for the second stage, whereas the computational complexity is significantly reduced. The performance loss of the proposed suboptimal detection method may depend on the signal constellation. For the considered 8 QAM constellation, the proposed detection method achieves near-ML detection performance.

The generalized detection approach can be used with arbitrary multidimensional signal constellations. Signal constellations are usually constructed with the D-fold Cartesian product of conventional two-dimensional signal constellations, i.e. each active transmit antenna employs a conventional complex-valued signal constellation. In Section 3.2, we have shown that multidimensional signal constellations, based on the densest known lattices, outperform the aforementioned construction in terms of error performance on the AWGN channel. In Section 5.2, we discuss signal constellations based on the Hurwitz integers and the application to GMSM transmission schemes. With the proposed construction method for the signal constellations, cardinalities which have a power of two can be created. As discussed in the first part of this chapter, with conventional two-stage detection schemes, the complexity of the symbol detection increases rapidly as more active transmit antennas are used. A complexity reduction can be obtained with the symbol detection scheme proposed in the first part of this chapter. This method combines equalization and a list-decoding approach. This technique can be generalized to lattice-based multidimensional signal constellations. We have demonstrated this generalization for four-dimensional Hurwitz constellations. The presented simulation results indicate that near-ML performance is achieved, whereas the computational complexity is significantly reduced. For large signal constellations, the Hurwitz constellations improve the performance of GMSM compared with conventional QAM or PSK constellations.

The proposed detection method and partitioning technique may also be used with $N_a > 2$. Therefore, symbols of the signal constellations are drawn from a suitable lattice according to the number of active antennas, e.g., for $N_a = 4$ the E_8 lattice [CS99] can be used. Similar to Hurwitz integers, all the coordinates of E_8 are either integers or half-integers. Hence, a similar partitioning is possible. However, the search for well-performing multidimensional signal constellations for more than two active antennas is still open.

In the third part of this chapter, we have discussed the application of OMEC codes $\mathcal{C}$ as multidimensional signal constellations, i.e. $\mathcal{X} = \mathcal{C}$, for GMSM. Such codes were used in Section 4.2 [RSF20] for SPM [LSL+19] with a single active transmit antenna. The simulation results presented in Section 4.2.3 [RSF20] demonstrate that the space-time codes based on codes over Gaussian and Eisenstein integers outperform conventional SPM. However, these results are based on ML decoding and the computational complexity of ML decoding increases exponentially with the code dimension. Here, we have proposed a novel transmission scheme for GMSM based OMEC codes over Gaussian or Eisenstein integers. This approach is enables a suboptimal list-decoding strategy, where in contrast to the previously discussed detection methods, the symbol is detected first and the antenna pattern afterward. The proposed GMSM scheme with suboptimal detection can outperform conventional GMSM schemes based on QAM or PSK constellations with ML detection. The simulation results show that the proposed method can outperform conventional GMSM with respect to decoding performance, detection complexity, and spectral efficiency.

Conclusions

6

In this thesis, signal constellations with algebraic properties have been investigated. Special focus was given for the application of these signal constellations for spatial modulation transmission. All considered spatial modulation variants are wireless MIMO-transmission techniques, where not all transmit antennas are active at the same time.

In the first part of this thesis, we have considered spatial modulation with a single active antenna. In this case, the complex-valued Gaussian- and Eisenstein-integer fields are directly applicable as signal constellation. These constellations enable simple yet powerful codes, the so-called one Mannheim error correcting codes [Hub94a, Hub94b]. With SM, the zero element is reserved to indicate inactive transmit antennas. Hence, the application of one Mannheim error-correcting code-based signal constellations in spatial modulation transmission requires some adaptation. We have shown that codes over Gaussian and Eisenstein integers outperform the repetition codes used in the original proposal in spatial permutation modulation [LSL+19]. The results are based on maximum-likelihood detection. However, higher spectral efficiencies would require suboptimal detection approaches. For the uncoded spatial modulation transmission, we have developed a low-complexity detection method that achieves near-optimum performance.

In the second part of the thesis, we have considered multidimensional signal constellations which are applicable with generalized multistream spatial modulation. With GMSM, multidimensional signal constellations are usually based on the D-dimensional fold Cartesian product of one conventional complex-valued signal constellation like QAM or PSK. Here we have developed two alternative approaches.

We have considered four-dimensional signal constellations with elements from the set of Lipschitz and Hurwitz integers. These constellations are useful for

D. B. Rohweder, *Signal Constellations with Algebraic Properties and their Application in Spatial Modulation Transmission Schemes*, Schriftenreihe der Institute für Systemdynamik (ISD) und optische Systeme (IOS), https://doi.org/10.1007/978-3-658-37114-2_6

generalized multistream spatial modulation with two active antennas. The Hurwitz constellations are based on the densest known lattices from the literature [CS99]. We have shown that these constellations can result in a performance gain, when compared with the Cartesian product of a conventional constellation. Moreover, we have proposed two suboptimal detection techniques for generalized multistream spatial modulation. One can be used with arbitrary multidimensional signal constellations. The second approach considers lattice-based signal constellations and exploits the dependencies over the dimensions. For Hurwitz integers, the detection complexity is reduced to the case when a conventional two-dimensional signal constellation with suboptimal detection is used. Moreover, for equal spectral efficiencies, the Hurwitz integers with suboptimal detection show lower error rates than the conventional QAM signal constellations with maximum-likelihood detection.

Finally, we have considered one Mannheim error correcting codes as multidimensional signal constellation. These codes enable an efficient suboptimum decoding method with near-optimum performance. For generalized multistream spatial modulation, this enables a new detection approach. The most common suboptimal detection procedure with spatial modulation is based on two stages: The first stage estimates the active transmit antenna(s) and the second stage estimates the transmitted symbol(s). We have developed a detection approach based on linear equalization of the channel matrix which allows independent detection of symbols. This approach reduces the complexity for the conventional two-stage detection approach with only a small performance loss. Combined with the suboptimal decoding of one Mannheim error correcting codes, it enables an efficient decoding method that is an alternative to the conventional two-stage detection. The simulation results for one Mannheim error correcting codes show that the proposed method can outperform conventional generalized multistream spatial modulation with respect to decoding performance, detection complexity, and spectral efficiency.

Bibliography

[ADP14] Y. Acar, H. Doğati, and E. Panayırcı. Iterative channel estimation for spatial modulation systems over fast fading channels. In *2014 22nd Signal Processing and Communications Applications Conference (SIU)*, pages 578–581, 2014.

[Agr21] E. Agrell. Database of sphere packings, Feb. 2021.

[BAPP12] E. Basar, U. Aygolu, E. Panayirci, and H. V. Poor. Performance of spatial modulation in the presence of channel estimation errors. *IEEE Communications Letters*, 16(2):176–179, 2012.

[Bos99] Martin Bossert. *Channel coding for telecommunications*. Wiley, 1999.

[BX16] T. Balmahoon and H. Xu. Low-complexity EDAS and low-complexity detection scheme for MPSK spatial modulation. *IET Communications*, 10(14):1752–1757, 2016.

[CS99] J.H. Conway and N.J.A. Sloane. *Sphere Packings, Lattices and Groups*. Springer Verlag, New York, Berlin, 3rd edition, 1999.

[CS03] John Horton Conway and Derek Alan Smith. *On quaternions and octonions: their geometry, arithmetic, and symmetry*. A. K. Peters Ltd., 2003.

[CSSS15] Chien-Chun Cheng, H. Sari, S. Sezginer, and Y.T. Su. Enhanced spatial modulation with multiple signal constellations. *IEEE Transactions on Communications*, 63(6):2237–2248, June 2015.

[DM98a] M. C. Davey and D. MacKay. Low-density parity check codes over GF(q). *IEEE Communications Letters*, 2(6):165–167, 1998.

[DM98b] M. C. Davey and D. J. C. MacKay. Low density parity check codes over GF(q). In *1998 Information Theory Workshop (Cat. No.98EX131)*, pages 70–71, June 1998.

[dPBZB12] N. di Pietro, J. J. Boutros, G. Zémor, and L. Brunel. Integer low-density lattices based on construction A. In *2012 IEEE Information Theory Workshop*, pages 422–426, Sept. 2012.

[FAZ09] M. M. U. Faiz, S. Al-Ghadhban, and A. Zerguine. Recursive least-squares adaptive channel estimation for spatial modulation systems. In *2009 IEEE 9th Malaysia International Conference on Communications (MICC)*, pages 785–788, 2009.

[FBW14] Dong Fang, A. Burr, and Yi Wang. Eisenstein integer based multi-dimensional coded modulation for physical-layer network coding over f4 in the two-way

D. B. Rohweder, *Signal Constellations with Algebraic Properties and their Application in Spatial Modulation Transmission Schemes*, Schriftenreihe der Institute für Systemdynamik (ISD) und optische Systeme (IOS), https://doi.org/10.1007/978-3-658-37114-2

relay channels. In *European Conference on Networks and Communications (EuCNC)*, pages 1–5, June 2014.

[FGS13] J. Freudenberger, F. Ghaboussi, and S. Shavgulidze. New coding techniques for codes over Gaussian integers. *IEEE Transactions on Communications*, 61(8):3114–3124, Aug. 2013.

[Fos96] G. J. Foschini. Layered space-time architecture for wireless communication in a fading environment when using multi-element antennas. *Bell Labs Technical Journal*, 1(2):41–59, 1996.

[FRS18] J. Freudenberger, D. Rohweder, and S. Shavgulidze. Generalized multistream spatial modulation with signal constellations based on Hurwitz integers and low-complexity detection. *IEEE Wireless Communications Letters*, 7(3):412–415, June 2018.

[FS15a] J. Freudenberger and S. Shavgulidze. New four-dimensional signal constellations from Lipschitz integers for transmission over the Gaussian channel. *IEEE Transactions on Communications*, 63(7):2420–2427, July 2015.

[FS15b] Juergen Freudenberger and Sergo Shavgulidze. New signal constellations for coding over Lipschitz integers. In *Proceedings of 10th International ITG Conference on Systems, Communications and Coding (SCC 2015)*, pages 1–6, Feb. 2015.

[FS17] J. Freudenberger and S. Shavgulidze. Signal constellations based on Eisenstein integers for generalized spatial modulation. *IEEE Communications Letters*, 21(3):556–559, Mar. 2017.

[FSFF20] F. Frey, S. Stern, J. K. Fischer, and R. F. H. Fischer. Two-stage coded modulation for Hurwitz constellations in fiber-optical communications. *Journal of Lightwave Technology*, 38(12):3135–3146, 2020.

[FSS14] J. Freudenberger, J. Spinner, and S. Shavgulidze. Set partitioning of Gaussian integer constellations and its application to two-dimensional interleaving. *IET Communications*, 8(8):1336–1346, May 2014.

[FW89] G.D. Forney and Lee-Fang Wei. Multidimensional constellations. I. Introduction, figures of merit, and generalized cross constellations. *Selected Areas in Communications, IEEE Journal on*, 7(6):877–892, Aug. 1989.

[Gal63] R. G. Gallager. *Low-Density Parity-Check Codes*. M.I.T. Press, Cambridge, Massachusetts, 1963.

[Gal68] R. G. Gallager. *Information Theory And Reliable Communication*. John Wiley & Sons Inc, New York, NY, USA, 1968.

[GH14] M. Güzeltepe and Olof Heden. Perfect Mannheim, Lipschitz and Hurwitz weight codes. *Math. Commun.*, 19:253–276, 2014.

[GL12] Gene H. Golub and Charles F. Van Loan. *Matrix Computations*. JHU Press, 2012.

[Gü13] M. Güzeltepe. Codes over Hurwitz integers. *Discrete Mathematics*, 313:704–714, 2013.

[Gü18] M. Güzeltepe. On some perfect codes over Hurwitz integers. *Mathematical Advances in Pure and Applied Sciences*, 1:39–45, 2018.

[Hua17] Y. Huang. Lattice index codes from algebraic number fields. *IEEE Transactions on Information Theory*, 63(4):2098–2112, 2017.

[Hub94a] K Huber. Codes over Eisenstein-Jacobi integers. *Contemporary Mathematics*, pages 165–179, Jan. 1994.

[Hub94b] K Huber. Codes over Gaussian integers. *IEEE Transactions on Information Theory*, pages 207–216, 1994.

[JGS08] J. Jeganathan, A. Ghrayeb, and L. Szczecinski. Spatial modulation: optimal detection and performance analysis. *IEEE Communications Letters*, 12(8):545–547, Aug. 2008.

[JKM00] Hui Jin, Aamod Khandekar, and Robert McEliece. Irregular repeat accumulate codes. In *2nd International Symposium on Turbo Codes and Related Topics*, September 2000.

[KA10] M. Karlsson and E. Agrell. Four-dimensional optimized constellations for coherent optical transmission systems. In *Optical Communication (ECOC), 2010 36th European Conference and Exhibition on*, pages 1–6, Sept. 2010.

[KMI+10] H. Kostadinov, H. Morita, N. Iijima, A. J. Han Vinck, and N. Manev. Decoding of integer codes and their application to coded modulation. *IEICE Trans. Fundamentals*, E93-A(7):1363 –1370, July 2010.

[KRAK12] D. Kapetanovic, F. Rusek, T. E. Abrudan, and V. Koivunen. Construction of minimum Euclidean distance MIMO precoders and their lattice classifications. *IEEE Transactions on Signal Processing*, 60(8):4470–4474, 2012.

[LC04] Shu Lin and Daniel J. Costello. *Error Control Coding. Upper Saddle River*, NJ: Prentice-Hall, 2004.

[Loe91] H.-A Loeliger. Signal sets matched to groups. *IEEE Transactions on Information Theory*, 37(6):1675–1682, Nov. 1991.

[LSL+19] I. Lai, J. Shih, C. Lee, H. Tu, J. Chi, J. Wu, and Y. Huang. Spatial permutation modulation for multiple-input multiple-output (MIMO) systems. *IEEE Access*, 7:68206–68218, May 2019.

[LWL15] C. T. Lin, W. R. Wu, and C. Y. Liu. Low-complexity ML detectors for generalized spatial modulation systems. *IEEE Transactions on Communications*, 63(11):4214–4230, Nov. 2015.

[MBG07] C. Martinez, R. Beivide, and E. Gabidulin. Perfect codes for metrics induced by circulant graphs. *IEEE Transactions on Information Theory*, 53(9):3042–3052, Sept. 2007.

[MHAY06] R. Mesleh, H. Haas, C. W. Ahn, and S. Yun. Spatial modulation – A new low-complexity spectral efficiency enhancing technique. In *in Proc. CHINACOM*, pages 1–5, Oct. 2006.

[MHS+08] R. Y. Mesleh, H. Haas, S. Sinanovic, C. W. Ahn, and S. Yun. Spatial modulation. *IEEE Transactions on Vehicular Technology*, 57(4):2228–2241, July 2008.

[NDPNV13] K. Ntontin, M. Di Renzo, A. Perez-Neira, and C. Verikoukis. Performance analysis of multistream spatial modulation with maximum-likelihood detection. In *IEEE Global Communications Conference (GLOBECOM)*, pages 1590–1594, Dec. 2013.

[NHV15] L. Natarajan, Y. Hong, and E. Viterbo. Lattice index coding. *IEEE Transactions on Information Theory*, 61(12):6505–6525, 2015.

[NXQ11] N. R. Naidoo, H. J. Xu, and T. A. M. Quazi. Spatial modulation: optimal detector asymptotic performance and multiple-stage detection. *IET Communications*, 5(10):1368–1376, July 2011.

[PA03] J. Porath and T. Aulin. Design of multidimensional signal constellations. *IEE Proceedings-Communications*, 150(5):317–323, Oct. 2003.

[PC93] S.S. Pietrobon and D.J. Costello. Trellis coding with multidimensional QAM signal sets. *IEEE Transactions on Information Theory*, 39(2):325–336, Mar. 1993.

[PDL+90] S.S. Pietrobon, R.-H. Deng, A Lafanechere, G. Ungerboeck, and D.J. Costello. Trellis-coded multidimensional phase modulation. *IEEE Transactions on Information Theory*, 36(1):63–89, Jan. 1990.

[PIF+01] T. Pires da Nobrega Neto, J. C. Interlando, O. M. Favareto, M. Elia, and R. Palazzo. Lattice constellations and codes from quadratic number fields. *IEEE Transactions on Information Theory*, 47(4):1514–1527, May 2001.

[Pro07] Proakis. *Digital Communications 5th Edition*. McGraw Hill, 2007.

[PX13] N. Pillay and H. Xu. Comments on ", signal vector based detection scheme for spatial modulation". *IEEE Communications Letters*, 17(1):2–3, Jan. 2013.

[RCO+17] R. Rath, D. Clausen, S. Ohlendorf, S. Pachnicke, and W. Rosenkranz. Tomlinson-Harashima precoding for dispersion uncompensated PAM-4 transmission with direct-detection. *Journal of Lightwave Technology*, 35(18):3909–3917, September 2017.

[RFS18] © 2018 IEEE. Reprinted, with permission, from D. Rohweder, J. Freudenberger, and S. Shavgulidze. Low-density parity-check codes over finite Gaussian integer fields. In *2018 IEEE International Symposium on Information Theory (ISIT)*, pages 481–485, June 2018.

[RSF+19] © 2019 IEEE. Reprinted, with permission, from D. Rohweder, S. Stern, R. F. H. Fischer, S. Shavgulidze, and J. Freudenberger. Low-complexity detection for generalized multistream spatial modulation. In *20th International Workshop on Signal Processing Advances in Wireless Communications (SPAWC)*, pages 1–5, July 2019.

[RSF20] © 2020 IEEE. Reprinted, with permission, from D. Rohweder, S. Shavgulidze, and J. Freudenberger. Codes over Gaussian integers for spatial modulation. In *24th International ITG Workshop on Smart Antennas (WSA)*, pages 1–6, Feb. 2020.

[SF05] D. Sridhara and T. E. Fuja. Ldpc codes over rings for psk modulation. *IEEE Transactions on Information Theory*, 51(9):3209–3220, Sept. 2005.

[SF15] S. Stern and R. F. H. Fischer. Lattice-reduction-aided preequalization over algebraic signal constellations. In *9th International Conference on Signal Processing and Communication Systems (ICSPCS)*, pages 1–10, 2015.

[SF18] S. Stern and R. F. H. Fischer. Quaternion-valued multi-user MIMO transmission via dual-polarized antennas and QLLL reduction. In *25th International Conference on Telecommunications (ICT)*, pages 63–69, June 2018.

[SFFF19] S. Stern, F. Frey, J. K. Fischer, and R. F. H. Fischer. Two-stage dimension-wise coded modulation for four-dimensional Hurwitz-integer constellations. In *12th International ITG Conference on Systems, Communications and Coding (SCC)*, pages 197–202, February 2019.

[SH12] S. Sugiura and L. Hanzo. Effects of channel estimation on spatial modulation. *IEEE Signal Processing Letters*, 19(12):805–808, 2012.

[Ste19] Sebastian Stern. *Advanced equalization and coded-modulation strategies for multiple-input/multiple-output systems*. PhD thesis, Ulm University, May 2019.

[SYHS13] Q. T. Sun, J. Yuan, T. Huang, and K. W. Shum. Lattice network codes based on Eisenstein integers. *IEEE Transactions on Communications*, 61(7):2713–2725, July 2013.

[THBN15] N. E. Tunali, Y. C. Huang, J. J. Boutros, and K. R. Narayanan. Lattices over Eisenstein integers for compute-and-forward. *IEEE Transactions on Information Theory*, 61(10):5306–5321, Oct. 2015.

[UR08] R. Urbanke and T. Richardson. *Modern coding theory*. Cambridge University Press, 2008.

[USS19] N. Ugrelidze, S. Shavgulidze, and M. Sordia. New generalized multistream spatial modulation for wireless communications. In *Proceedings of 11th Wireless Days (WD-2019)*, pages 1–7, Apr. 2019.

[USS20] Nodar Ugrelidze, Sergo Shavgulidze, and Mariam Sordia. New four-dimensional signal constellations construction. *IET Communications*, pages 1–7, 2020.

[WCDH14] X. Wu, H. Claussen, M. Di Renzo, and H. Haas. Channel estimation for spatial modulation. *IEEE Transactions on Communications*, 62(12):4362–4372, 2014.

[WFH99] U. Wachsmann, R.F.H. Fischer, and J.B. Huber. Multilevel codes: theoretical concepts and practical design rules. *IEEE Transactions on Information Theory*, 45(5):1361–1391, July 1999.

[WJS12a] J. Wang, S. Jia, and J. Song. Generalised spatial modulation system with multiple active transmit antennas and low complexity detection scheme. *IEEE Transactions on Wireless Communications*, 11(4):1605–1615, Apr. 2012.

[WJS12b] J. Wang, S. Jia, and J. Song. Signal vector based detection scheme for spatial modulation. IEEE Communications Letters, 16(1):19–21, Jan. 2012.

[XYD+14] Y. Xiao, Z. Yang, L. Dan, P. Yang, L. Yin, and W. Xiang. Low-complexity signal detection for generalized spatial modulation. *IEEE Communications Letters*, 18(3):403–406, Mar. 2014.

[YDX+15] P. Yang, M. Di Renzo, Y. Xiao, S. Li, and L. Hanzo. Design guidelines for spatial modulation. *IEEE Communications Surveys Tutorials*, 17(1):6–26, Firstquarter 2015.

[YSMH10] A. Younis, N. Serafimovski, R. Mesleh, and H. Haas. Generalised spatial modulation. In *44th Asilomar Conference on Signals, Systems and Computers*, pages 1498–1502, Nov. 2010.

Publications of the Author

[1] J. Freudenberger, M. Rajab, D. Rohweder, and M. Safieh. A codec architecture for the compression of short data blocks. *Journal of Circuits, Systems, and Computers (JCSC)*, 27(2):1–17, Feb. 2018.

[2] © 2018 IEEE. Reprinted, with permission, from J. Freudenberger, D. Rohweder, and S. Shavgulidze. Generalized multistream spatial modulation with signal constellations based on Hurwitz integers and low-complexity detection. *IEEE Wireless Communications Letters*, 7(3), 412–415, June 2018.

[3] J. Freudenberger, D. Rohweder, and S. Shavgulidze. Low-complexity detection for spatial modulation. In *12th International ITG Conference on Systems, Communications and Coding (SCC)*, pages 1–5, Feb. 2019.

[4] © 2018 IEEE. Reprinted, with permission, from D. Rohweder, J. Freudenberger, and S. Shavgulidze. Low-density parity-check codes over finite Gaussian integer fields. In *2018 IEEE International Symposium on Information Theory (ISIT)*, pages 481–485, June 2018.

[5] © 2020 IEEE. Reprinted, with permission, from D. Rohweder, P. Oleschak, S. Shavgulidze, and J. Freudenberger. Generalized multistream spatial modulation based on one mannheim error correcting codes and their low-complexity detection. In *10th IEEE International Conference of Consumer Technology (ICCE)*, Oct. 2020.

[6] D. Rohweder, S. Shavgulidze, and J. Freudenberger. Codes over Gaussian integers for spatial modulation. In *24th International ITG Workshop on Smart Antennas (WSA)*, pages 1–6, Feb. 2020.

[7] © 2019 IEEE. Reprinted, with permission, from D. Rohweder, S. Stern, R. F. H. Fischer, S. Shavgulidze, and J. Freudenberger. Low-complexity detection for generalized multistream spatial modulation. In *20th International Workshop on Signal Processing Advances in Wireless Communications (SPAWC)*, pages 1–5, July 2019.

[8] D. Rohweder, S. Stern, R. F. H. Fischer, S. Shavgulidze, and J. Freudenberger. Low-complexity detection for multi-dimensional spatial modulation schemes. In *24th International ITG Workshop on Smart Antennas (WSA)*, pages 1–6, Feb. 2020.

[9] © 2021 IEEE. Reprinted, with permission, from D. Rohweder, S. Stern, R. F. H. Fischer, S. Shavgulidze, and J. Freudenberger. Four-dimensional hurwitz signal constellations, set partitioning, detection, and multilevel coding. *IEEE Transactions on Communications*, 69(8), 5079–5090, Aug. 2021.

[10] M. Safieh, D. Rohweder, and J. Freudenberger. Implementierung einer speichereffizienten Huffman-decodierung. In *58th Multi Project Chip (MPC) Workshop*, volume 58/59, pages 27–31, July 2018.

[11] J. Spinner, D. Rohweder, and J. Freudenberger. Soft input decoder for high-rate generalised concatenated codes. *IET Circuits, Devices and Systems*, 12(4), 432–438, July 2018.

[12] S. Stern, D. Rohweder, J. Freudenberger, and R. F. H. Fischer. Binary multilevel coding over Eisenstein integers for MIMO broadcast transmission. In *23rd International ITG Workshop on Smart Antennas (WSA)*, pages 1–8, Apr. 2019.

[13] S. Stern, D. Rohweder, J. Freudenberger, and R. F. H. Fischer. Multilevel coding over Eisenstein integers with ternary codes. In *12th International ITG Conference on Systems, Communications and Coding (SCC)*, pages 1–6, Feb. 2019.